ICE
MICRODYNAMICS

ICE
MICRODYNAMICS

PAO K. WANG

Department of Atmospheric and Oceanic Sciences
University of Wisconsin-Madison
Madison, Wisconsin

ACADEMIC PRESS
An imprint of Elsevier Science

Amsterdam   Boston   London   New York   Oxford   Paris
San Diego   San Francisco   Singapore   Sydney   Tokyo

This volume is a paperback reprint of Pao K. Wang's article that appears
in Volume 45 of *Advances in Geophysics* (R. Dmowska and B. Saltzman, eds.).

This book is printed on acid-free paper. ∞

Academic Press
An imprint of Elsevier Science.
525 B Street, Suite 1900, San Diego, California 92101-4495, USA
http://www.academicpress.com

Academic Press
84 Theobalds Road, London WC1X 8RR, UK
http://www.academicpress.com

Library of Congress Catalog Card Number: 20022002107825

International Standard Book Number: 0-12-734603-1

PRINTED IN THE UNITED STATES OF AMERICA
02 03   04   05   06   07   MM   9 8   7   6   5   4   3   2   1

To Professor Dr. Hans R. Pruppacher

CONTENTS

Shape and Microdynamics of Ice Particles and Their Effects in Cirrus Clouds

PAO K. WANG

PREFACE

I spent a semester of sabbatical leave at MIT in the fall of 1997. On my way back to Madison, I visited Yale University's Professor Barry Saltzman, who encouraged me to summarize my research about ice particles in clouds into a monograph. This volume is the result of just such an endeavor.

Ice particles are present both in the lower atmosphere (during hailstorms and snowstorms for example) and in upper atmospheric clouds such as cirrus and the upper parts of cumulonimbus. Earlier studies of clouds strongly emphasized the physics of liquid processes, as ice processes were then not very well understood. On the observational side, the difficulty lay in accessing the higher parts of clouds where ice particles are located. On the theoretical side, the main difficulty was that attending the more complex ice particle shapes. My own research in Wisconsin in the past two decades has focused largely on theoretical studies of ice particles in clouds, especially the mathematical description of their shapes, their diffusional and collisional growth rates, and their influence on cloud development. All these seem to be related, in one way or another, to their hydrodynamic properties. That is why I use the word "microdynamics" in the title to distinguish from the usual word "microphysics," which tends to cover all microscale cloud processes.

This volume mainly consists of my research works on this subject, most of which have been published in referred scientific journals; however, some unpublished results are also included. These latter include some results that are too detailed for journal articles, insights that occurred to me after a paper was published, and excerpts from some of my students' Ph.D theses. Formal journal articles are being prepared to disseminate these theses excerpts. This monograph is not intended to be a comprehensive treatment of ice microdynamics.

The historical notes in Section 1 are not formally related to microdynamics; rather, they represent a passage of human discovery of ice crystal behavior in the atmosphere. Thus I feel it is of some interest and not totally out of place here. Joseph Needham had discussed some of them, but others—from my own reading—are probably presented in English for the first time. The second half of Section 1 briefly introduces the later sections. Sections 2 through 4 are concerned with the microdynamic behavior of individual ice particles—their size and shape, their hydrodynamics, the diffusion of vapor around them, and their collision efficiencies with small droplets. Section 5 focuses on the impact of ice microdynamics on the scavenging of aerosol particles, a process that may play an important role in the upper tropospheric chemical transport. Finally, the subject of Section 6 is cirrus cloud development, where ice microdynamics assumes a central role. A cirrus model equipped with cloud microphysics and radiation packages is used for the cirrus study.

I am grateful to the late Professor Barry Saltzman for his kind invitation, which made this monograph possible, and for the hospitality that he (and Professor Ron Smith) gave me during my short stay in New Haven. (Barry passed away on February 2, 2001.) I am indebted to my colleague Dr. Bob Schlesinger, who tirelessly read the manuscript and made innumerable corrections and comments that resulted in great improvements. (And, from what Bob has done, I finally understand what the word "meticulous" really means!) Of course, I bear the responsibility for any mistakes in this volume. I would also like to acknowledge my current and former students who have dedicated much of their youthful energy into these research. I also thank two long-time friends, Dr. Andy Heymsfield and Professor Ken Beard, for many discussions about my research and helpful comments on the manuscript.

The results presented here were achieved mainly under the generous support of research grants from National Science Foundation. I especially acknowledge Drs. Ron Taylor and Rod Rogers, the respective former and current Directors of the Physical Meteorology Program, and Dr. Steve Nelson, Director of the Mesoscale Dynamic Meteorology Program. All have been very supportive of my research in ice processes in clouds. I also acknowledge the support of the U.S. Environmental Protection Agency for my work in the aerosol scavenging studies. The Alexander von Humboldt Foundation of Germany also provided me with a generous Humboldt Award that allowed me to perform part of my research in Germany. The American Meteorological Society has the most generous policy of allowing researchers to use materials from its journals for their publications. Dr. Ed Eloranta also kindly provided me a beautiful lidar picture of cirrus clouds. I also thank Ms. Erin Lunders, who patiently typed the first draft, and Dr. Frank Cynar and Mr. Paul Gottehrer of Academic Press, who provided much assistance that shaped the final form of this monograph. Ms. Kelly Ricci carefully typed the final draft.

Finally, I dedicate this volume to Professor Hans R. Pruppacher, from whom I learned my basic lessons on ice physics in clouds and whose monumental book *Microphysics of Clouds and Precipitation* is a source that never fails to inspire me when I need some cloud information. Hans is a model scientist, a teacher, and a friend. He and his wife, Monica, introduced our family to the charm of Europe and gave us the most pleasant experience during our stay in Mainz, Germany, in 1993.

Pao K. Wang
Madison, Wisconsin

SHAPE AND MICRODYNAMICS OF ICE PARTICLES AND THEIR EFFECTS IN CIRRUS CLOUDS

Pao K. Wang

Department of Atmospheric and Oceanic Sciences
University of Wisconsin-Madison
1225 W. Dayton Street
Madison, Wisconsin 53706

1. Ice Particles in the Atmosphere

1.1. Ice Particles—A Personal Perspective

Except for those who live in the polar regions, most people associate ice particles only with winter and cold climates. The reality is that ice particles are more ubiquitous than they realize. Even in the tropics in summer, ice particles may roam in the skies. Some of them may be small ice crystals in cirrus clouds, sometimes too thin to be visualized. Others may be snow crystals, graupel, and hailstones in vigorously developing cumulonimbus clouds. The only reason that common tropical dwellers do not see ice particles (unless they live on or near high mountains) is that they melt completely when they fall through the very warm air in the lower troposphere there. Even the upper parts of tropical storms—hurricanes (typhoons)—contain many ice particles.

Dwellers in middle and high latitudes are, of course, more familiar with ice particles, mostly as snow in winter but occasionally as hail in severe thunderstorms in summer. Here the air below the cloud base is sometimes cold and dry enough to allow hailstones to survive to the ground, causing great grief to farmers whose fruits or other crops may be badly damaged. There are other forms of ice particles as well, such as the frost on the grass in a cold morning in the fall and the menacing freezing rain in the early winter. They are "atmospheric" ice particles in a sense because their origins are in the atmosphere.

The Chinese ideograms representing ice and snow have been in existence for at least 3000 years. The earliest existing records are those engraved on the oracle bones of the Shang Dynasty (ca. 1000 B.C.). Aside from their usual usage in weather, they are often used to describe the color white, for they are probably the whitest things one can see in nature. In *Chuang-Tzu,* a book attributed to (and possibly written, at least partially, by) the great philosopher Chuang Chou (ca. 4th century B.C.), a passage reads:

> In the Mt. Miao-Gu-Ye there lives a goddess whose flesh and skin are as white as ice and snow and who looks like a graceful and beautiful virgin.

So it appears that Snow White is not necessarily just a Western stereotypical beauty.

The author was born and grew up in Taiwan, a tropical country where natural ice particles exist only in high mountains (according to some statistics, some 60 peaks there are taller than 10,000 feet!). Until about 20 years ago these high mountains were not readily accessible to the public (except for those diehard mountain climbers, of course). As a result, most people living in Taiwan have never seen snow, even though the usage of "snow" is very common in the language. When I was a high school student, I hiked in high mountains in wintertime and surely saw ice packed by the roadside, but I never saw actual snowfall. I remember that one year a little flurry occurred in Chi-Sin-Shan [Seven Star Mountain], a mountainous area close to Taipei. After the news was reported in the media, thousands of people jammed the highway, racing to the area to see the snow. According to one of my friends who went to see it, the snows "look like a layer of thin flour spreading on grass." Years later when I came to Madison, Wisconsin and was wading one winter morning in knee-high snow after a major blizzard, I couldn't help but burst into a big fit of laughter recalling the "snow flours" in Taipei.

The most impressive aspect of an ice crystal is its extremely elegant geometric design. For a layman knowing little about crystallography, it is hard to believe that something produced by natural processes can be so beautiful, intricate, and apparently made to great precision. Who is not impressed when looking at the album of snow crystal pictures photographed by that venerated Vermont farmer Bentley (Bentley and Humphreys, 1931)? The one that strikes me most is the picture of a capped ice column that looks like a perfectly made pillar, complete with graceful engravings, taken from an ancient Greek temple. Not only are they beautiful and elegant, but the designs are so complex and detailed that one gets an impression, as an old saying has it, that no two snowflakes are exactly alike. Even if they look the same on the surface, there must be some subtle differences in details. Thus came an interesting encounter between the news media and me. One day a reporter from a major local newspaper called me up wanting me to verify whether it is true that no two snowflakes are exactly alike. I wanted to be as precise as possible and replied, "It depends how detailed you really want to get. Two simple hexagonal ice plates without any visible internal designs may look exactly alike. But if you go down to the molecular level, you are bound to find some differences." The next day the news article read, "Wang confirms that it is indeed true that no two snowflakes look exactly alike." So much for my "scientific" answer!

In fact, as reported in *Science Now* (March 1995), the notion that no two snow crystals are identical was disproved in 1988 when National Center for Atmospheric

Research scientist Nancy Knight, who was examining samples collected at 6 km over Wisconsin for a cloud-climatology study, found two thick hollow columnar crystals that apparently were Siamese twins.

1.2. Some Historical Notes on the Knowledge of Ice Particles in Ancient China

It is of some interest to review the history of our scientific knowledge about ice particles. Chapter 1 of the book by Pruppacher and Klett (1997) has provided such a review for Western history. However, relatively few such works have been done for the case of China, where there is also a long tradition of meteorological observations. An exhaustive review of the observations would not be possible at present, but in the following few paragraphs I would just like to mention a few interesting observations about ice particles in ancient China.

1.2.1. The Hexagonal Shape of Snowflakes

According to Needham and Lu (1961), the first European to write something about the shape of snow crystals was Albertus Magnus (ca. A.D. 1260). He thought that snow crystals were star-shaped, but he also seemed to believe that such regular forms of snow fell only in February and March. In 1555, the Scandinavian bishop Olaus Magnus wrote that the snow crystals could have shapes like crescents, arrows, nail-shaped objects, bells, and one like a human hand. It was only in 1591 that Thomas Hariot correctly recognized the hexagonal nature of snow crystals (unpublished private manuscript). It is therefore quite an impressive feat that in 135 B.C., Han Ying, a Chinese scholar of the Western Han Dynasty, wrote in his book *Han Shih Wai Chuan [Moral Discourses Illustrating the Han Text of the Book of Odes]* about the hexagonal shape of snow crystals.The passage reads as follows:

> Flowers of plants and trees are generally five-pointed, but those of snow, which are called *ying*, are always six-pointed.

Han did not say anything about how the observations were made, but Needham and Lu wondered whether some kind of magnifying lens was used because this kind of discovery would imply fine-scale examination. Ever since Han's work was published, the six-sided nature of snowflakes has been a household term for Chinese scholars, and numerous writings, especially poems, allude to this fact. Needham and Lu mentioned one poem written by Hsiao Tung, a sixth-century prince of Liang Empire:

> The ruddy clouds float in the four quarters of the cerulean sky.
> And the white snowflakes show forth their six-petaled flowers.

When I was about five years old, my father showed me a popular textbook used by children of ancient China (up to the beginning of 20th century) when they first became students. In this book, *You Hsue Gu Shih Chong Ling [A Fine Jade Forest of Stories for Beginning Students]*, there is a sentence of a verse saying, "The flying of snow flakes, which are six-pointed, is an auspicious omen for good harvest." Snow is considered auspicious because it is believed to kill insect eggs/larvae, but it is also clear that the hexagonal shape of snow crystals is common knowledge among the masses.

Although there was no crystallographic explanation of the hexagonal nature of the crystals, the Chinese attempted to explain this six-sidedness by the explain-all principle of the Yin-Yang and Wu-Shing [Five Elements] theory. It was believed that these were the two opposing forces operating in the whole universe, and everything in this world is produced from their interaction. Yang is said to relate to male or positive aspects, whereas yin is said to relate to female or negative aspects of nature. Curiously, yang is also associated with odd numbers, whereas yin is associated with even numbers. In principle, everything in the universe can be categorized as belonging to either yin or yang. Not surprisingly, water is considered to possess the quality of yin, a female attribute, and hence is associated with an even number. For an unexplained reason, the number six was assigned to the element Water in the ancient time. So ancient Chinese scholars took this semimagic principle to "explain" why snow crystals are six-pointed. An example was given by the great 12th-century Chinese medieval philosopher Chu Hsi, who wrote, "Six generated from Earth is the perfect number of Water, so as snow is water condensed into crystal flowers, these are always six-pointed." In another writing in *Chu Tzu Chuan Shu [Collected Writings of Master Chu]*, Chi Hsi also tried to explain the formation of snow in a slightly more detailed way:

> The reason why "flowers" or crystals of snow are six-pointed is because they are only sleet split open by violent winds they must be six-pointed. Just so, if you throw a lump of mud on the ground, it splashes into radiating angular petal-like form. Now six is a yin number, and tai-yin-hsuan-jing-shi (selenite crystal) is also six-pointed, with sharp prismatic angular edges. Everything is due to the number inherent in Nature.

The ancient Chinese seemed content to accept that the six-pointed shape of snow is a natural fact, and no further study on its hexagonal nature was done.

1.2.2. Protection of Crops from the Cold

One concern about ice particles is the possibility that they may cause damage because of their coldness. This is especially relevant for the case of frost, a form of ice particles, which may severely damage fruits or other crops. Since China has been mainly an agricultural country through most of its history and can be cold, it is only natural to expect that the Chinese have paid some attention to the protection of crops from cold damage. In a Wei Dynasty agricultural book, *Chi Min Yao Shu [Essential Technologies for Common People]*, Jia Si-shie (early 6th century), we find two accounts concerning the protection of crops from frost damage:

(1) In the book written by Fan Sheng-Chih [A.D. 1st century] it says, "In planting rice, it is common to watch during the midnights during the period at about 80 or 90 days after the summer solstice. There may be frosts looking like white dews descending from the sky. [If this happens] at dawn, have two persons holding the two ends of a long rope and facing each other to scrape the frosts and dews from the rice plants. Continue to do it until the sunrise. This will prevent the damage of all crops."

To my knowledge, this is the only account ever mentioned in the literature of this direct scraping technique for frosts. In addition to the direct effect, two persons walking back and forth with a long rope over the crop field may also help stir up the air and cause the cold air below and the warm air above to mix, hence raising the temperatures near the crops. If this is the case, then the technique is essentially similar in principle to the wind machines used in some modern orchards to prevent frosts, but predated them by nearly 1900 years.

(2) If frost occurs during the blossoming season of fruit trees, the trees will bear no fruits. [To prevent this] one must first stock weeds and animal feces beforehand in the garden. When the sky clears up right after the rain and the north wind is cold and intense, frost will occur during that night. [If this happens] ignite the fuel to make thick smokes and the frost will be prevented.

This technique is exactly like the "orchard burner" technique of frost prevention used in modern orchards, but predated it by about 1400 years.

1.2.3. Observations of Frost-Free and Dew-Free Situations in High Mountains

Chu Hsi, the aforementioned philosopher of the Sung Dynasty, wrote in *Chu Tzu Yu Lei [Analects of Master Chu]* that

Frosts are merely frozen dewdrops and snow is merely frozen raindrops. Ancient folks related that dews are produced by the vapors of the stars and the moon. This statement is wrong. Nowadays on high mountains there is no dew even if the sky is clear. So the vapors that form dewdrops must come from below.

In this passage Chu Hsi recognized that if the dews (and hence the frosts, because back then people believed that frosts were frozen dews) came from the condensation of vapors descended from the stars and moon, then they should be abundant in high mountains that presumably are closer to those celestial bodies than low-lying ground is. Yet his observations indicated that that was not the case. He correctly concluded that the water vapor necessary for dew formation must come from below.

In the same book, Chu Hsi also mentioned that "on top of high mountains there is no frost or dew." This observation is probably not valid for all mountains because, in my own memory, there were dews early on summer morning in mountains of Taiwan. But this may be the case in the drier mountain areas where Chu Hsi lived. On the other hand, he also observed that there is snow on high mountains. Of particular interest is the way Chu Hsi reasoned about these phenomena:

> Someone asked, "What is the reason that there is no frost and dew on high mountains?" Answer: As you go up higher, the air becomes clearer and windier. Even if you have some moisture to begin with, the winds would have diluted them and prevented them from condensation. On the other hand, snow is caused by the freezing of raindrops when they are chilled. Hence one would see snow first in high and cold places.

In this passage Chu Hsi made a clear distinction between the formation of snow and that of frost even though they are both ice particles. His idea that snow is caused by freezing of moisture in the air aloft, and frost by freezing of moisture in low levels, is basically sound. Of course, what Chu said about snow formation is only true for the riming process whereas the more complete picture of snow formation had to wait until the work of Bergeron and Findeisen in the 20th century. But consider that Chu Hsi was a 13th-century observer!

1.3. A Brief Summary of the Following Sections

In the following few sections, the studies done by the author's research group concerning the microdynamic aspects of ice particles will be summarized. The main theme of these studies focuses on ice processes in the atmosphere, and the main effect emphasized here is the microdynamics of ice particles—that is, the effect of their motions. Ice microdynamics is important because ice particles move in a viscous medium, namely, air. The motions cause complicated flow fields around the falling ice particles, influencing their growth rates and hence the overall development of the cloud. Other processes in the atmosphere, such as transport of trace chemicals and radiative transfer, are certainly influenced greatly by the cloud development.

Section 2 details our studies on the mathematical descriptions of ice particles (mainly hexagonal snow crystals and conical graupel and hail). I start with the two-dimensional expressions and then expand to the three-dimensional expressions. The 3-D expressions allow the representation of hexagonal columns of finite lengths, hexagonal plates of finite thickness, and conical particles with elliptical cross sections. Such expressions should simplify the calculations of the physical properties (heat and mass transfer, flow fields, and scattering properties) of these particles. They may also be used for quantitative descriptions of the size distributions of these particles.

Section 3 deals with the numerical calculation of flow fields around falling ice particles. The main importance of such studies is that the flow fields determine the growth rates of ice particles for both diffusional and collisional modes. In the former, the flow fields determine the ventilation coefficients of vapor fluxes toward ice particles due to their falling motion. In the latter, the flow fields determine the collision efficiencies between particles. Owing to computational difficulties, we were able to calculate the flow fields for falling ice particles only at low and intermediate Reynolds numbers. But in these calculations we used more realistic shapes (finite-length columns, hexagonal plates, and broad-branch crystals with finite thickness) than in all other previous studies; moreover, we studied the case of unsteady flow to a certain extent. We hope to perform calculations of flow fields for larger falling ice hydrometeors such as large snowflakes, graupel, and hail in the future. The tasks will be very challenging because the flow will be fully unsteady, the particles' fall attitudes will be complicated, and viscous effects will have to be considered.

The numerical determinations of the ventilation coefficients for falling ice particles and of the collision efficiencies of ice crystals collecting supercooled cloud droplets are the subjects of Section 4. Both endeavors require the knowledge of flow fields around falling ice crystals, and these fields are taken from the results presented in Section 3. Using the ventilation coefficients presented here, we can determine the diffusional growth and sublimation rates of ice crystals falling in air (subject to limitations on the Reynolds number range, of course). This is especially useful in the modeling of cirrus clouds, in which the ice particles are mainly small to medium-sized crystals; we have also done some preliminary work in this aspect, as is discussed in Section 6. The collision efficiencies determined here can be applied to the calculation of riming rates that are important to the growth of graupel and hail.

From Section 2 through Section 4, we focus on the physical properties of individual ice particles. In Section 5, we look into the details of microdynamic processes by which aerosol particles are removed by ice. Here two complementary models of aerosol collection, one valid for smaller particles and the other for larger particles, are developed. The combined results from these two models give collection efficiencies over the whole particle size spectrum. Finally, these theoretical results are compared with some experimental measurements.

Section 6 is devoted to the study of cirrus clouds. It is increasingly accepted that the distribution of cirrus clouds has very important repercussions on global climate. One of the crucial questions is how long a cirrus cloud can survive under a special environmental condition. We constructed a two-dimensional cirrus cloud model with detailed cloud microphysics and radiation, and studied the development of cirrus clouds under four representative conditions representing stable and unstable atmospheric conditions in tropical and midlatitude regions. The results we obtained show that there are indeed complicated interactions among cloud microphysics, cloud dynamics, and radiative fields.

2. MATHEMATICAL DESCRIPTIONS OF ICE PARTICLE SIZE AND SHAPE

2.1. Size Distribution versus Size–Shape Distributions

It is clear that, except for some small frozen cloud droplets, ice particles in clouds are basically nonspherical. This nonsphericity causes many difficulties in the quantitative treatment of physical processes involving ice particles. For example, it is

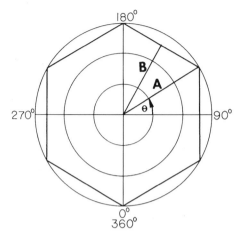

FIG. 2.1. The "size" of a hexagonal plate ice crystal is a function of the polar angle in this two-dimensional projection.

known that the optical properties of clouds depend not only on the size and concentration of their constituent particles, but also on the shape of these particles. Now, suppose we have a set of measured optical properties (absorption, scattering, etc.) for various parts of a cloud. How do we express the relation between these properties and the "aggregation state" of the cloud particles? For a cloud composed of only spherical particles, this would not be a problem because only the distribution of one variable, either the radius or the diameter, is needed to completely describe this aggregation state. This is known as the *size distribution* (see, for example, Chapter 2 of Pruppacher and Klett, 1997). Therefore, optical properties of such a cloud can be expressed as a function of the size distribution of cloud particles.

But for an ice cloud composed of nonspherical ice particles, it is no longer so simple. In this case, even the very definition of "size" is ambiguous. What, for example, is the size of a hexagonal ice plate? From Figure 2.1 it is clear that the size here is a function of the angle θ and is not a unique number. Moreover, even if two plates have the same horizontal dimensions, they could have different thicknesses. These problems exist for other ice crystal shapes as well. The problem becomes even messier when we are dealing with an ensemble of ice particles of different habits (shapes) and sizes. In short, the aggregation state of an ensemble of nonspherical particles cannot be described by the size distribution alone but has to be specified by the shape–size distributions.

A conventional method of classifying ice particle shapes was given by Magono and Lee (1966). The Magono–Lee classification is qualitative and thus useful in the descriptive categorization, but it is inadequate in providing quantitative information about the crystals. What we need here are some simple mathematical expressions that can describe both the size and shape of an ice particle so that its geometric properties (surface area, cross-sectional area, volume, etc.) can be calculated easily. "Simple" here means that these expressions should contain only a few adjustable parameters (and the fewer the better) while reproducing the correct size and essential shape of the ice particle. If this can be done, such expressions can be used to describe the shape and size of ice particles, and the distributions of these parameters can serve to characterize the shape and size characteristics of the ensemble.

2.2. Mathematical Expression Describing the Two-Dimensional Shapes of Hexagonal Ice Crystals

Let us start with the mathematical expression that describes the two-dimensional cross-sectional shapes of ice crystals. Only hexagonal ice crystals are treated, as this is the most common shape of ice crystals. This has been done by Wang and Denzer (1983) and Wang (1987, 1997).

Certainly there is more than one way to describe the hexagonal shapes, but the method discussed here is probably the simplest and also allows easy classification. Since hexagonality is a form of periodicity, it is intuitively appealing to invoke sine and cosine functions. We can choose the sine function here without loss of generality. For hexagonal shapes, we note that the function $f(\theta) = [\sin^2(3\theta)]$ produces six peaks in the range $0 \leq \theta \leq 2\pi$ and is therefore suitable for describing the hexagonal shape of snow crystals (see Fig. 2.2). To modulate the amplitude, we need only multiply $f(\theta)$ by a constant. On the other hand, the width of the peaks can be modulated by raising $f(\theta)$ to a positive power b, where b can be any positive number. The peaks are broad when b is small and are narrow when b is large because $0 \leq \sin^2(3\theta) \leq 1$.

Based on this idea, called the successive modification of simple shapes (SMOSS), we can change a simple shape into another shape by successively modifying it by some mathematical functions. The following expression is the one that *smosses* a circle ($r = c$) into a polygonal shape (in polar coordinates):

$$r = a[\sin^2(n\theta)]^b + c \qquad (2.1)$$

where r and θ are the radial and angular coordinates, respectively; and a, b, c, and n are adjustable parameters to fit the shape and size of the ice crystals. The ranges of these parameters are as follows:

a from $-c$ to ∞ (amplitude parameter)

b from 0 to ∞ (width parameter)

c from 0 to ∞ (center size parameter)

n $0, \frac{1}{2}, 1, \frac{3}{2}, 2, \ldots$ (polygonality parameter)

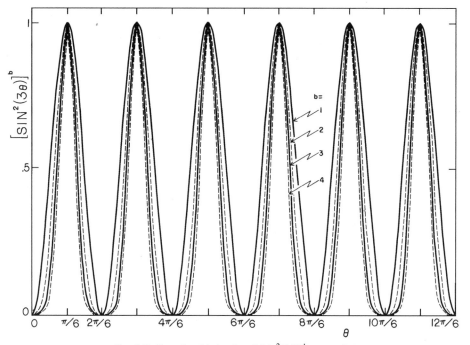

FIG. 2.2. Functional behavior of $[\sin^2(3\theta)]^b$ versus θ.

The number of sides of the polygon generated by (2.1) is $2n$ because of the square of the sine function. For the characterization of ice crystals, $n = 3$ since most ice crystals are hexagonal.

Examples of ice crystal shapes generated by Eq. (2.1) will be given later.

2.2.1. Determination of the Parameters a, b, and c for an Ice Crystal

Given that a hexagonal ice crystal shape can be described by Eq. (2.1), how do we determine the parameters $a, b,$ and c if we are given a real ice crystal sample? This can be done in reference to Figure 2.3. The method described here is more suitable for fitting ice crystals with rounded branches. [Here we will consider only idealized symmetric crystals. Real crystals often have unequal branch lengths (dendrites) or are missing branches, and sometimes have surface features (pits); but these are not of concern here.]

From the geometry of Figure 2.3, it is easy to see that

$$c_1 = a\left(\tfrac{1}{2}\right)^b + c \tag{2.2}$$

$$c_2 = a + c \tag{2.3}$$

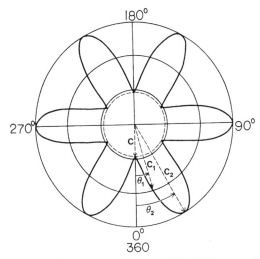

FIG. 2.3. Definition of c, c_1, and c_2 for Eqs. (2.2) and (2.3). This hexagonal crystal is generated by Eq. (2.1) by setting $a = 5.13$, $b = 1$, $c = 2.49$ [Bentley and Humphreys (1962), p. 143, (3,1)].

where c_1 and c_2 are the radial lengths at $\theta = \pi/12$ and $\pi/6$, respectively. From (2.2) and (2.3) we get

$$b = \ln[(c_2 - c)/(c_1 - c)]/\ln 2 \qquad (2.4)$$

Equations (2.3) and (2.4) provide all the calculations needed to fit Eq. (2.1) to a real snow crystal. The steps are as follows:

1. Measure c, c_1, and c_2.
2. Determine a from Eq. (2.3).
3. Determine b from Eq. (2.4).

The area enclosed by Eq. (2.1) can be determined from the following formula:

$$A = \frac{1}{2} \int_0^{2\pi} r^2 \, d\theta \qquad (2.5)$$

where r in the integral, of course, is to be replaced by the expression on the RHS of (2.1). For an ice crystal of thickness h, the volume is simply $V = Ah$.

2.3. Approximating an Exact Hexagonal Plate

The simplest shape of hexagonal ice crystals is the hexagonal plate whose ideal cross section is an exact hexagon. It is of interest to see if we can use Eq. (2.1)

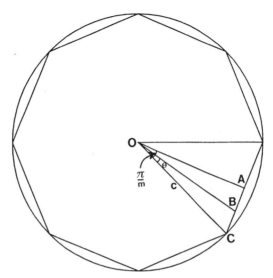

FIG. 2.4. Definitions of notations used in Section 2.4.

to approximate an exact hexagon. It turns out that we can do this to a fairly good degree (Wang, 1987).

Let us examine the general scheme of using Eq. (2.1) to approximate exact polygons (of which the hexagon is just a special case). Figure 2.4 shows a sector of an exact polygon which is circumscribed by the circle with center at O and radius C. If the polygon is m-sided, then the angle $\angle COA$ is π/m. The length of the line segment $\overline{OA}$ is

$$\overline{OA} = C \cos(\pi/m) \tag{2.6}$$

The length of line segment OB is therefore

$$\overline{OB} = \frac{\overline{OA}}{\cos[(\pi/m) - \theta]} = \frac{C \cos(\pi/m)}{\cos[(\pi/m) - \theta]} \tag{2.7}$$

so that

$$\overline{OB} - C = C \left\{ \frac{\cos(\pi/m)}{\cos[(\pi/m) - \theta]} - 1 \right\} \tag{2.8}$$

The RHS of (2.8) is the amount to be subtracted from the circle in order to produce the polygon, and exactly the amount that should be approximated by the first term on the RHS of Eq. (2.1). In order to achieve a close approximation, this term should be made to generate a value as closely as possible to the RHS of (2.8) for all θ. Clearly we have to let $m = 2n$ if (2.1) is to be used. In order to determine the optimum value of a, we note that the "wave" height at $\theta = \pi/m = \pi/2n$ generated by (2.1) is a maximum (i.e., the amplitude):

$$a[\sin^2(n\theta)]^b = a[\sin^2(\pi/2)]^b = a \tag{2.9}$$

at $\theta = \pi/2n$, which, in the most ideal situation, should be equal to the RHS of Eq. (2.8); i.e.,

$$a = c\left[\frac{\cos(\pi/2n)}{\cos[(\pi/2n) - (\pi/2n)]} - 1\right] = c[\cos(\pi/2n) - 1] \tag{2.10}$$

or

$$a/c = \cos(\pi/2n) - 1 \tag{2.11}$$

The above equation determines the optimum a for given values of c and n. To determine the optimal value of b, we substitute Eq. (2.10) in the first term of (2.1) and require that the resultant expression approximates (2.8) for all θ, i.e.,

$$c\left[\cos\left(\frac{\pi}{2n}\right) - 1\right][\sin^2(n\theta)]^b \approx c\left[\frac{\cos(\pi/2n)}{\cos[(\pi/2n) - \theta]} - 1\right] \tag{2.12}$$

or

$$[\sin^2(n\theta)]^b \approx \left[\frac{\cos(\pi/2n)}{\cos[(\pi/2n) - \theta]} - 1\right] \Big/ \left[\cos\left(\frac{\pi}{2n}\right) - 1\right] \tag{2.13}$$

for all θ. The simplest way of determining the optimum width parameter b is perhaps the least-squares method, which requires that the sum of the squares of the differences between the LHS and RHS of (2.13) be a minimum for all θ. This can be easily done by iteration. Table 2.1 shows the values of a/c and b for generating approximated exact polygons with $m(= 2n)$ sides. For a hexagon, the values turn out to be $a/c = -0.1339$ and $b = 0.397$. Figure 2.5 shows a few examples of the polygons generated by Eq. (2.1) using values of a/c and b as specified in Table 2.1. It is seen that, while the cases for the triangle and square are less than satisfactory, the approximation looks rather good for $m \geq 5$. [It turns out that a square can be generated nearly exactly by Eq. (2.14).]

Equation (2.1) can produce many other shapes in addition to those described above. For details, see Wang (1987).

2.4. Two-Dimensional Characterization of an Ensemble of Planar Hexagonal Ice Crystals

We have seen that Eq. (2.1) can be used to fit various shapes of planar hexagonal ice crystals. As indicated earlier, the distributions of the parameters a, b, and c can serve to characterize the shapes and sizes of an ensemble of such ice crystals, with $n = 3$ as explained before.

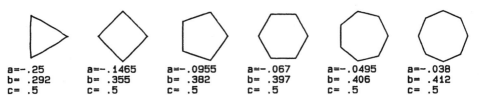

a=-.25   a=-.1465   a=-.0955   a=-.067   a=-.0495   a=-.038
b= .292   b= .355   b= .382   b= .397   b= .406   b= .412
c= .5   c= .5   c= .5   c= .5   c= .5   c= .5

Fig. 2.5. Approximated exact polygons using Eq. (2.11) with values of a, b, and c given in Table I.

In addition to characterizing actual ice or snow crystal samples, Eq. (2.1) can be used to generate model samples of ice crystals in clouds. The latter application may be particularly useful for cloud modeling work. Recent cloud models often contain detailed cloud microphysics to describe the growth of, or interaction between, various kinds of cloud and precipitation particles (Cotton and Anthes, 1989; Johnson et al., 1993, 1994). Until now most models have implemented only generic categories of ice particles without specifying their habits. Yet it is known that crystal habits do influence the diffusional and collisional growth rates (Pruppacher and Klett, 1997; Wang and Ji, 1992) and also the radiative properties of clouds (see Sec. 6). To include crystal habit features in the model quantitatively, we can use Eq. (2.1) to generate ensembles of ice crystals. First, the cloud model would determine the temperature and saturation ratio of a particular cloud region. Then an

TABLE 2.1 VALUES OF a/c AND b FOR GENERATING
APPROXIMATED EXACT POLYGONS WITH $2n$ SIDES

n	a/c	b
1.5	−0.5000	0.292
2	−0.2929	0.355
2.5	−0.1910	0.382
3	−0.1339	0.397
3.5	−0.0990	0.406
4	−0.0761	0.412
4.5	−0.0603	0.416
5	−0.0489	0.418
5.5	−0.0405	0.421
6	−0.0341	0.422
6.5	−0.0291	0.423
7	−0.0251	0.424
7.5	−0.0219	0.425
8	−0.0192	0.426
8.5	−0.0170	0.426
9	−0.0152	0.427
9.5	−0.0136	0.428
10	−0.0123	0.428

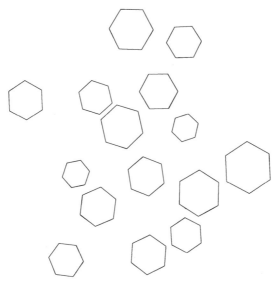

FIG. 2.6. A simulated sample of 15 hexagonal ice plates generated by Eq. (2.1).

appropriate crystal habit can be selected according to this temperature and saturation ratio. Suppose only one crystal habit, say, a hexagonal plate, is to be assigned
to a certain cloud region, in which case the parameter b is fixed. Next, the amount
of excess moisture to be converted into ice is distributed into specified spectra of
a and c. The distributions of a, b, and c so specified determine not only the size
but also the shape of ice crystals in this region.

Figure 2.6 shows such a hypothetical ensemble of hexagonal plates whose distribution of a, b, and c are given in Figure 2.7. Numerical values of a, b, and c are
given in Table 2.2. Since these are congruent shapes, the b distribution assumes
a single value, with $b = 0.397$ as determined in the last section. In this example,
the a and c spectra are chosen to resemble gamma-type distributions. Note that
in this sample, the value of a/c is fixed, so that the a and c distributions are not
independent. In fact, to characterize this ice crystal sample, it is only necessary to
specify either an a or c distribution and note that $b = 0.397$.

The case of mixed-habit ice crystals is more complicated. Again we create a
hypothetical ice crystal sample whose (a, b, c) values are given in Table 2.3. The
appearance of this sample is shown in Figure 2.8 and the distributions of the
a, b, and c parameters are given in Figure 2.9. Again, the a and c distributions
are chosen to be quasi-gamma type, whereas the b distribution is now bimodal.
Obviously the dominant habit here is broad-branch crystals. Figure 2.9 appears to
be a "reasonable" ice crystal sample.

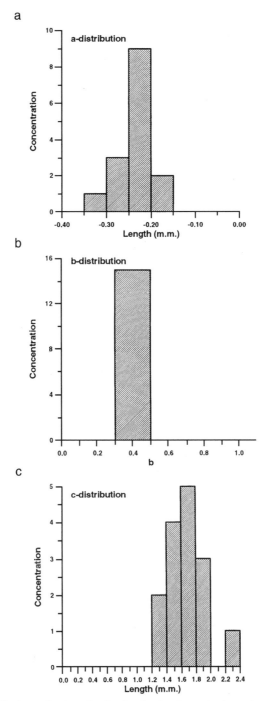

FIG. 2.7. Distributions of a, b, and c for the simulated sample shown in Figure 2.6. In this sample, $b = 0.397$.

TABLE 2.2 VALUES OF AMPLITUDE (a) AND CENTER
SIZE (c) PARAMETERS AND THE CONCENTRATION OF
HEXAGONAL ICE CRYSTALS SHOWN IN FIGURE 2.6

a	c	Concentration
0.18	1.35	2
0.21	1.55	4
0.23	1.75	5
0.27	2.00	3
0.30	2.25	1

Note: In this sample, $n = 3$, $b = 0.397$.

At this point, it is useful to sound a cautionary note about the nature of shape–size
distributions. While it is true that a well-documented ice crystal ensemble will lead
to a unique set of shape–size distributions, the reverse is not necessarily true. For
a sample of ice crystals with a single habit, one can always reconstruct the crystals
in Figure 2.6 from the information in Figure 2.7. But this is not true for mixed-
habit ice crystal samples. For example, an ensemble of crystals whose shape–size
parameters are given in Table 2.4 will have the same a, b, and c distributions also
represented by Figure 2.9. Only the orders of the a and b values are randomly
switched. Yet the shapes of these ice crystals, shown in Figure 2.10, are not the
same as those in Figure 2.8. However, the two ensembles look rather similar, so

TABLE 2.3 VALUES OF a, b, AND c PARAMETERS
OF THE SIMULATED ICE CRYSTAL SAMPLE AS
SHOWN IN FIGURE 2.8

a	b	c
−0.68	2.0	1.16
−1.97	3.5	2.45
−1.85	2.1	2.77
−1.00	3.3	1.05
−1.24	1.7	1.75
−1.54	3.6	1.65
−2.31	5.2	2.44
−1.17	1.6	1.85
−1.15	2.7	2.05
−1.57	0.9	1.96
−0.35	10.0	1.66
−0.15	0.6	1.97
−0.75	10.0	1.96
−1.05	5.0	1.54
−1.75	50.0	1.80

FIG. 2.8. A simulated sample of 15 ice crystals generated by Eq. (2.1) with the distributions of a, b, and c given in Figure 2.9.

they may not have significantly different physical properties. Whether this is true for all occasions is unclear at present.

2.5. Mathematical Expressions Describing the Three-Dimensional Shapes of Ice Crystals

The previous few sections dealt mainly with mathematical descriptions of the two-dimensional cross-sectional shapes of hexagonal ice crystals. Real ice crystals are, of course, three-dimensional, and a complete description of their geometric characteristics needs three-dimensional mathematical formulas, such as those developed below.

2.5.1. Three-Dimensional Mathematical Expression Describing Planar Hexagonal Crystals

Since Eq. (2.1) can generate the 2-D cross-sectional shapes of hexagonal crystals, we need to consider how to represent the thickness of the crystal and combine it into the 2-D shape formula. To do this, we start with the simpler question of how to find a mathematical expression to describe a finite circular cylinder. If this can be done, we can use Eq. (2.1) to transform the circular cross section of the cylinder into a hexagon so that we have a 3-D hexagonal crystal. If the length is small, we have a plate; if the length is large, we have a hexagonal column.

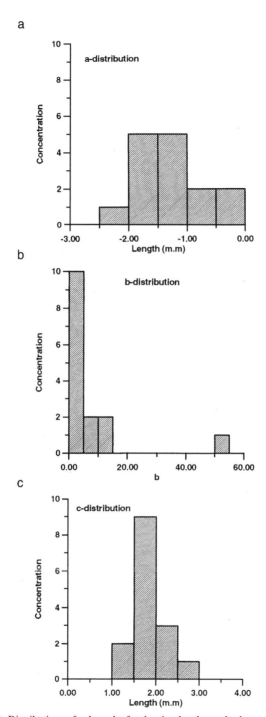

FIG. 2.9. Distributions of a, b, and c for the simulated sample shown in Figure 2.8.

PAO K. WANG

TABLE 2.4   VALUES OF *a*, *b*, AND *c* PARAMETERS
OF THE SIMULATED ICE CRYSTAL SAMPLE AS SHOWN
IN FIGURE 2.10

a	b	c
−0.68	1.6	1.80
−1.97	2.7	2.05
−1.85	0.9	2.44
−1.00	10.0	1.05
−1.24	0.6	1.75
−1.54	10.0	1.65
−2.31	5.0	2.77
−1.17	50.0	1.66
−1.15	2.0	2.45
−1.57	3.5	1.96
−0.35	2.1	1.85
−0.15	3.3	1.97
−0.75	1.7	1.96
−1.05	3.6	1.16
−1.75	5.2	1.80

FIG. 2.10. A simulated sample of 15 ice crystals generated by Eq. (2.1) with the values of *a*, *b*, and *c* given in Table 2.4.

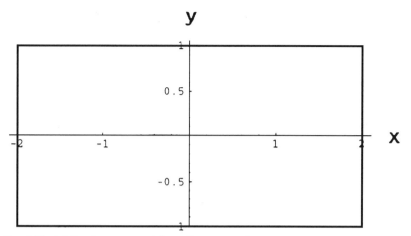

FIG. 2.11. A 2-D columnar crystal (in reality, a rectangle) generated by Eq. (2.14) with $a = 2$ and $c = 1$.

We note that the cross section of a finite circular cylinder through its axis is a rectangle. If we can find an equation closely approximating the rectangle, then rotating the resulting curve along its long axis will yield a close approximation of a finite circular column.

Wang (1997, 1999) has given the equation describing a rectangle in Cartesian coordinates:

$$\frac{x^2}{a^2} + \frac{z^2}{c^2}\left(1 + \varepsilon - \frac{x^2}{a^2}\right) = 1 \tag{2.14}$$

where x and z are the common Cartesian coordinates; a and c are the half-lengths in the x and z direction, respectively; and ε is an adjustable positive parameter that can be set as small as we wish (but never equal to zero)[1] to closely fit the sharp corners of a rectangle. Larger values of ε would result in more "rounded" corners, while smaller values of ε produce sharper corners. For regular purposes, it may be sufficient to set $\varepsilon = 10^{-5}$. Indeed, Figure 2.11 shows a "rectangle" generated by Eq. (2.14) with $a = 2$, $c = 1$, and $\varepsilon = 10^{-5}$. Since ε is finite, the rectangle represented by (2.14) is differentiable at every point.

It is now easy to generate the right circular cylinder from the rectangle as represented by (2.14) by simply rotating it about an axis. If we now let the length be in the vertical z direction, then the cylinder is given by the following expression (see Fig. 2.12)

$$\left(\frac{x^2 + y^2}{a^2}\right)\left(1 + \varepsilon - \frac{z^2}{c^2}\right) + \frac{z^2}{c^2} = 1 \tag{2.15}$$

[1]If $\varepsilon = 0$, then discontinuities occur at $x = \pm a$, and Eq. (2.14) represents two parallel line segments.

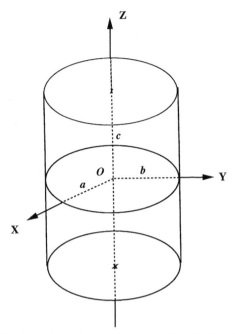

FIG. 2.12. A circular cylinder generated by Eq. (2.15).

This is the expression representing a right cylinder to a high degree of precision if ε is set small enough. The cross section of the cylinder in (2.15) is very close to a circle of radius a. The "equator" of the cylinder is slightly constricted but only by a fractional amount of the order ε.

To turn this cylinder into a hexagonal column of length c, all we need to do is to transform the circular cross section into a hexagon. This is the form given by Eq. (2.11). Changing the symbols therein to go with the change to a 3-D situation, we have

$$a \rightarrow a - A[\sin^2(3\varphi)]^B \qquad (2.16)$$

where the expression on the RHS of (2.16) is in 2-D polar coordinates and φ is the angular coordinate. A and B are adjustable parameters that change the shape of the cross section. Transforming Eq. (2.15) via (2.16) would then give a column of finite half-length c with hexagonal cross sections in mixed coordinates. Thus it is necessary either to transform (2.16) into Cartesian coordinates or to transform the final result into spherical coordinates. This is done in the following section.

If we apply the rectangular transformation (2.14) in Eq. (2.15), then we would obtain a rectangular column (or a cube if all sides are of equal length) instead of a circular column. This will not be discussed here.

Cartesian Coordinates Representation

We note that

$$\sin(3\varphi) = 3\sin\varphi\,\cos^2\varphi - \sin^3\varphi$$

so that

$$\sin^2(3\varphi) = \left[\frac{3(a\sin\varphi)(a^2\cos^2\varphi)}{a^3} - \frac{a^3\sin^3\varphi}{a^3}\right]^2$$

$$= \frac{y^2(3x^2 - y^2)^2}{(x^2 + y^2)^3} \tag{2.17}$$

where we have utilized the fact that on a circle of radius a in polar coordinates, we have

$$\begin{cases} x = a\cos\varphi \\ y = a\sin\varphi \end{cases} \tag{2.18}$$

Thus, by substituting (2.16) into (2.15) and changing the resulting equation into Cartesian form, the expression for a column of hexagonal cross section is given by

$$\frac{(x^2 + y^2)}{[a - A\{y^2(3x^2 - y^2)/(x^2 + y^2)^3\}^B]^2}\left(1 + \varepsilon - \frac{z^2}{c^2}\right) + \frac{z^2}{c^2} = 1 \tag{2.19}$$

Spherical Coordinates Representation

To express (2.19) in spherical coordinates, we use the conventional metrics

$$\begin{cases} x = r\sin\theta\cos\varphi \\ y = r\sin\theta\sin\varphi \\ z = r\cos\theta \end{cases} \tag{2.20}$$

where r, θ, and φ are radial, zenith angular, and azimuth angular coordinates, respectively. By substituting (2.20) into (2.19), we get

$$\frac{r^2\sin^2\theta}{\{a - A[\sin^2(3\varphi)]^B\}^2} = \left[\frac{1 - (r^2\cos^2\theta)/c^2}{1 + \varepsilon - (r^2\cos^2\theta)/c^2}\right] \tag{2.21}$$

which is the expression desired.

The parameters to be specified in order to generate a particle with hexagonal cross sections are a, c, A, and B (the parameter ε is considered to be preset). Once these four parameters are specified, both the size and the shape of

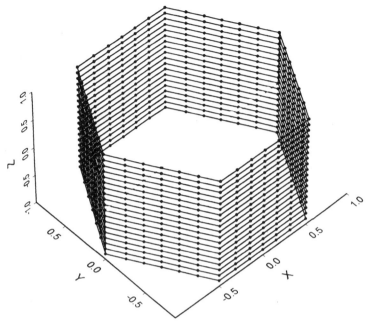

FIG. 2.13. A short hexagonal column generated by Eq. (2.21) with $a = 1$, $c = 1$, $A = 0.1339$, and $B = 0.397$. Dots represents data points computed from Eq. (2.21) and are connected by lines to show the prism surfaces of the simulated ice crystal. Note that Eq. (2.21) also generates points on the basal surfaces, which are not shown to avoid confusion.

the particle are completely fixed. The relative magnitudes of a and c determine whether the particle looks more nearly planar or columnar, according to whether $c > a$ or $a > c$, respectively. The parameters A and B determine the shape of the cross section, as explained in Wang (1987, 1997). Figures 2.13, 2.14, and 2.15 give three examples of the ice particles specified by (2.19) or (2.21). Figure 2.13 represents a hexagonal ice column while Figure 2.14 represents a hexagonal ice plate. The only difference between the two is the length parameter c. Figure 2.15 represents the shape of a broad-branch crystal of the same thickness as the ice plate in Figure 2.14, albeit of a different cross section. The reader is referred to two papers (Wang, 1987, 1997) for more detailed descriptions of various cross-sectional shapes and how to generate them. It is emphasized here that Eqs. (2.19) and (2.21) represent not only the prism surfaces but also the basal surfaces, which are not shown in Figures 2.13, 2.14, and 2.15 for the sake of clarity. The degree of flatness of both surface types is controlled by the parameter ε. The smaller ε is, the closer the prism and basal surfaces to real "planes."

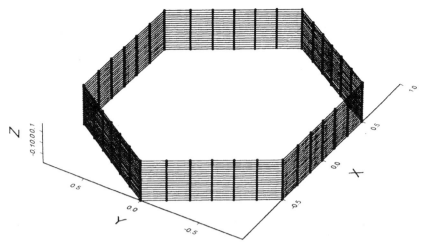

FIG. 2.14. A hexagonal plate generated by Eq. (2.21) with $a = 1$, $c = 0.2$, $A = 0.1339$, and $B = 0.397$.

The surface and cross-sectional areas and volumes of the particles generated by Eqs. (2.19) and (2.21) can be easily obtained. The method of calculating the cross-sectional area of the particle is given in Wang and Denzer (1983) and Wang (1987). The volume of the ice crystal is simply the cross-sectional area times its length c. The surface area is the sum of the basal planes ($= 2 \times$ cross-sectional

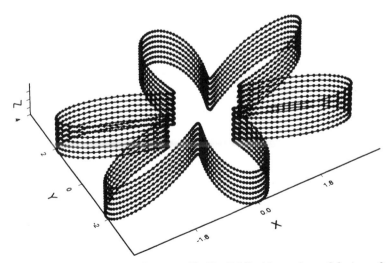

FIG. 2.15. A broad-branch crystal generated by Eq. (2.21) with $a = 1$, $c = 0.2$, $A = -3$, and $B = 1$.

area) plus the area of the prism surface (= length × the perimeter of the cross section). The perimeter of any shape represented by (2.16) can be determined by the integral

$$\int_0^{2\pi} r \, d\varphi$$

where r represents the expression on the RHS of (216).

2.5.2. Three-Dimensional Mathematical Expressions Describing Spatial Dendrites and Rosettes

Ice crystals in clouds are, of course, not all planar. What we have developed in the previous few sections may describe the basic shapes (while ignoring details such as the delicate dendritic branches), but they obviously fall short in describing many ice crystals whose characteristics are conspicuously three-dimensional, e.g., C2a, P5b, P7b, and CP2a in the classification of Magono and Lee (1966). Some of these crystals, e.g., bullets in cirrus clouds, may occur relatively often. Before we try to devise formulas to describe the three-dimensional ice crystals, it is useful to briefly review their structures.

Three-dimensional ice crystals are thought to grow from frozen drops and hence are usually polycrystalline. They are not well understood relative to single crystals, and only a handful of studies exist (see, for example, Hallett, 1964; Lee, 1972; Hobbs, 1976; Kobayashi et al., 1976a,b; Kikuchi and Uyeda, 1979; Furukawa, 1982). Unlike the single crystals treated in previous sections, these crystals are usually not polygonally symmetric. Even with a single type of crystal, such as a combination of bullets, the angles between different branches may differ widely in different samples. Figure 2.16 is an example of the frequency histograms of angles between the c axes of each component for the case of bullets (from Kobayashi et al., 1976a). Strangely, the other two more common types of three-dimensional crystals—the spatial dendrites and the radiating dendrites—both have a dominant angle frequency at 70° (Lee, 1972; Kobayashi et al., 1976; Kikuchi and Uyeda, 1979).

Although it may be possible to simulate rather closely the shape of these three-dimensional crystals, the ensuing mathematical expressions would be rather complicated, especially in reproducing the 70° angle, which is not evenly divisible by 360°. The purpose of introducing mathematical formulas here is to use them for estimating the bulk distributions of ice water contents and performing first-order calculation of some simple diffusional and radiative properties, and not for precise crystallographic investigations. Therefore, it is thought that simple formulas that produce polygonally symmetric shapes, instead of shapes with 70° angles, are probably sufficient. In addition, at this writing, it seems that the collection, preservation, and analysis of a large sample of three-dimensional crystals would

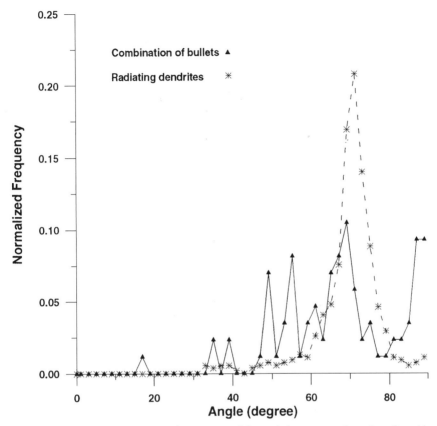

FIG. 2.16. Normalized frequency of occurrence of the angle between two branches of combinations of bullets and radiating dendrites. (Data adapted from Kobayashi *et al.,* 1976a.)

be a rather intractable effort, so it is probably impractical to try characterizing an actual sample. Perhaps, the holographic technique will eventually advance far enough so that such tasks can be easily and economically done. On the other hand, it is relatively easy to construct hypothetical samples with characteristics similar to actual samples. It is for this purpose that the formulas described below are designed.

With the above considerations in mind, it is possible to generalize the two-dimensional equation (2.1) to three-dimensions so as to simulate two more commonly observed three-dimensional ice crystals, namely, combinations of bullets and radiating dendrites. These expressions are described below.

Bullets are fairly common in cirrus clouds (see, for example, Heymsfield, 1975; Parungo, 1995). Cirrus clouds are known to have considerable influence on the radiative budget and hence the climate of the earth–atmosphere system. Therefore,

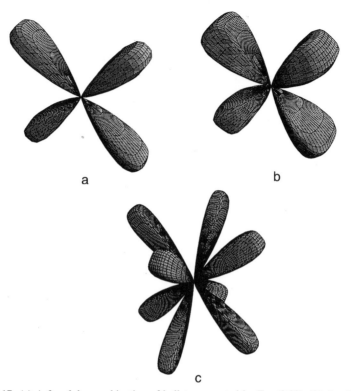

FIG. 2.17. (a) A four-lobe combination of bullets generated by Eq. (2.22). (b) Another example of a four-lobe combination of bullets generated by (2.22) except that $b = b' = 6$ in this case. (c) An example of an eight-lobe combination of bullets generated by $r = [1 - \cos(2\varphi)^4]^{20} [1 - \sin(2\theta)^4]^{20}$.

it is of significance to examine the expressions that can simulate such a shape. The following expression can be used to generate the combination of bullets:

$$r = \{a[\cos^2(m\theta)]^b + c\}^d \{a'[\sin^2(n\varphi)]^{b'} + c'\}^{d'} \tag{2.22}$$

This equation is simply Eq. (2.1) applied to both the θ and φ directions. Thus, it can be expected that it will produce shapes similar to those shown in previous sections when looking at cross sections for some specific values of θ or φ. The shape generated by this expression will have $2mn$ branches. For example, a four-branch combination of bullets as shown in Figure 2.17a can be generated by the following expression:

$$r = [1 - \cos^4(2\theta)]^{20}[1 - \sin^4(\varphi)]^{20} \tag{2.23}$$

where $m = 2$ and $n = 1$ here. The angle between branches is evidently 90°, which also occurs in nature as indicated by Figure 2.16. The widths of the branches are controlled by b and b' in Eq. (2.22). For example, if we change the values of $b = b' = 2$ in Eq. (2.22) to $b = b' = 3$, then the branches will look "fatter" as shown in Figure 2.17b. In both examples, the branches have relatively flat end surfaces that are close to the actual samples. However, there are cases where the end surfaces are capped plates that are not simulated here. The shape of each branch is not hexagonal, in contrast to real ice bullets. It is unclear at this point whether this really matters much in terms of the bulk radiative properties. It is also noted that hexagonal columns have hollows, as do bullets, but they are not considered here.

Figure 2.17c is an example of a bullet combination with 8 ($m = n = 2$) branches. The angle between branches is again 90°, but this time the branches are distributed in two mutually perpendicular planes. A "broken" combination, say, only the lower half, can be generated by selecting only the values of θ from $\pi/2$ to π.

The other habit to be considered here is the radiating dendrites. Again, it is impractical to simulate the intricate designs of each branch, but the basic shape of the crystal can be represented by the following expression:

$$r = a[\sin^2(m\theta)]^b[\sin^2(n\varphi)]^{b'} + c \qquad (2.24)$$

This is an obvious extension of Eq. (2.1) from 2-D to 3-D, and hence it is easy to see that it will produce a 3-D polygonally symmetric shape. Figure 2.18 is an example of radiating dendrites generated by Eq. (2.23) by assigning $a = 0.1$, $b = b' = 30$, and $c = 0.001$. The large value of b and b' are chosen in order to make the branches very thin.

FIG. 2.18. A radiating dendrite generated by Eq. (2.24) with $a = 0.1$, $b' = 30$, and $c = 0.001$.

Fitting observed 3-D crystals by the above expressions is analogous to the procedure for 2-D crystals, except the measurements may be difficult to perform as mentioned previously. The easiest way to do this is probably by measuring the 2-D projection of the crystals and then performing the fitting process for the 2-D crystals as described in Wang and Denzer (1983) and Wang (1987). Since the expressions given here generate polygonally symmetric shapes, it should be easy to obtain a 2-D projection.

2.6. Mathematical Expressions Describing Conical Hydrometeors

In the previous few sections we are mainly concerned with pristine ice crystal shapes. In this section, we treat conical hydrometeors.

There are two main kinds of conical hydrometeors: (a) conical graupel and hailstones and (b) large raindrops. Berge investigated a total of 1,920 hailstones and reported that 21% of them were conical while 41, 10, and 8 percent were oblate, spherical and prolate respectively (Battan, 1973). Although the fraction of conical hailstones was not the largest, it is large enough to deserve some special attention. Many graupel particles are also conical, as has been reported by Knight and Knight (1970), Locatelli and Hobbs (1974), and Hobbs (1976), among others. Falling raindrops of millimeter size usually have a "hamburger bun" shape with a more or less round top and flattened bottom (Pruppacher and Klett, 1997). Under certain conditions, such as in a strong vertical electric field, drops can be elongated to become pear-shaped (Pruppacher *et al.,* 1982; Richards and Dawson, 1971). They can also be thought of as conical particles.

Conical graupel and hailstones can be approximated by spherical sectors or the combination of a flat-based cone and a spherical cap, as has been done by Jayaweera and Mason (1965) and List and Schemenauer (1971). While these approximations are certainly workable and much has indeed been learned by making such assumptions, they also have some shortcomings. For example, these approximations all consist of two surfaces, namely, a conical surface and a spherical surface. This may complicate theoretical study of these particles because the two surfaces constitute a mixed boundary problem, which is usually more difficult to solve than a simple boundary. The shapes of falling raindrops have been investigated theoretically by Pruppacher and Pitter (1971), who used a cosine series to represent the shape of a deformed drop. Their method is useful when dealing with the detailed drop shape under the influence of certain forces. On the other hand, for some studies of physical properties such as the heat diffusion rates from such drops and the flow fields around them, it may be desirable to have a simpler mathematical function than a series to describe their shape.

Accordingly, we next present a single mathematical function that can describe the shape of conical graupel and hailstones as well as some deformed raindrops.

This function allows us to calculate fairly easily the volume, cross-sectional area, and the surface area of revolution of these hydrometeors. In addition, as in the case for pristine ice crystals, the parameters involved in this function may also serve as a method of classifying the dimensions of the hydrometeors.

2.6.1. Mathematical Formula—Two-Dimensional Cross Section

The mathematical function under consideration is

$$x = \pm a\sqrt{1 - z^2/C^2}\ \cos^{-1}(z/\lambda C) \tag{2.25}$$

where x and z are the respective horizontal and vertical coordinates of the surface (see Fig. 2.19), while a, C, and λ are parameters to be determined. The parameters a and C have dimensions of length whereas λ is a dimensionless number; C is one-half the length from the apex to the bottom along the z-axis, the center point being defined as the origin O, and a is defined in the following paragraph. We first note that

$$x = \pm a\sqrt{1 - z^2/C^2} \tag{2.26}$$

is the equation of an ellipse whose semi-axes in the x and z directions are a and C, respectively. Therefore, Eq. (2.25) can be thought of as an ellipse modulated

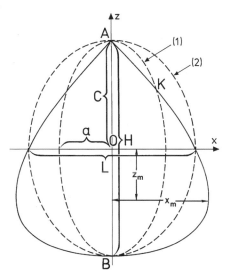

FIG. 2.19. Definitions of the coordinate system and various quantities appearing in Eq. (2.25). Solid curve is an axial cross section of a conical body. Dashed curves (1) and (2) are generating and limiting ellipses, respectively.

by an arccosine function. We may conveniently call the ellipse of Eq. (2.26) the *generating ellipse*. Therefore a is the horizontal semi-axis of the generating ellipse. The principal value of the arccosine function lies between 0 and π, i.e.,

$$0 \leq \cos^{-1}(z/\lambda C) \leq \pi \qquad (2.27)$$

Also since the absolute value of the cosine function cannot exceed 1 and $-C \geq z \geq C$, it is necessary that

$$\lambda \geq 1 \qquad (2.28)$$

We need not consider negative values of λ, since z can be either positive or negative. When $\lambda \to \infty$, $(z/\lambda C) = 0$, and

$$\cos^{-1}(z/\lambda C) = \cos^{-1}(0) = \pi/2 \qquad (2.29)$$

so that Eq. (2.25) becomes, in the limit,

$$x = \pm(\pi/2)a\sqrt{1 - z^2/C^2} \qquad (2.30)$$

which is the equation of an ellipse with horizontal and vertical semi-axes $\pi a/2$ and C, respectively. Since it represents the limit as $\lambda \to \infty$, it may be conveniently called the *limiting ellipse*. Curves representing Eqs. (2.25), (2.26), and (2.30) are shown in Figure 2.19. We see that Eq. (2.25) defines a pear-shaped curve. It will be shown later that Eq. (2.25) can approximate the shape of conical hydrometeors. The curve in Figure 2.19 represents an axial cross section only. To obtain a three-dimensional body, it is only necessary to rotate the curve about the z-axis.

2.6.2. Fitting Procedure and Examples

Various conical shapes can be obtained by changing parameters a, C and λ in Eq. (2.25). In the following we outline the steps that determine a, C, and λ for a real conical hydrometeor. We use a conical graupel as an example here.

Step 1: First we determine C. This is obtained by measuring the length H from the apex A to the bottom point B (refer to Fig. 2.19). C is one-half of this length, the center point O is defined as the origin, and the line of length H lies on the z-axis.

Step 2: Next we determine a. First draw the x-axis and measure the length L, as shown in Fig. 2.19. Of course, $z = 0$ along the x-axis. Therefore from Eq. (2.25) we have

$$L = 2x = 2a\sqrt{1 - z^2/C^2} \; \cos^{-1}(z/\lambda C)$$

which becomes, when $z = 0$

$$L = a\pi$$

so that

$$a = L/\pi \tag{2.31}$$

Step 3: Finally we have to determine λ. This can be divided into two cases:

Case 1: The conical curve intersects the generating ellipse
This is probably the most common case for conical graupel and hailstones. In this case the function

$$\cos^{-1}(z/\lambda C)$$

is smaller than 1 for some values (usually positive) of y. Right at the point of intersection (point K in Fig. 2.19), the value of the arccosine function is

$$\cos^{-1}(z_K/\lambda C) = 1$$

where z_K is the z-coordinate of point K. Thus

$$z_K/\lambda C = \cos(1) = 0.5403$$

Therefore

$$\lambda = z_K/0.5403C \tag{2.32}$$

Hence substituting the values of z_K and C determine the value of λ.

Note that the value of λ determined from Eq. (2.32) may be slightly less than 1 due to measuring errors or deviation from standard shape. In this case we have to set $\lambda = 1$ since, by definition, the argument of the arccosine function must satisfy

$$z_K/\lambda C \leq 1$$

However, this small correction does not produce a large deviation from the actual shape.

Case 2: The conical curve does not intersect the generating ellipse

In this case, $\cos^{-1}(z/\lambda C) \geq 1$ along the conical curve. We can determine the value of λ from its relation to the maximum value of the x-coordinate, x_m. In the following we only have to consider the half curve

$$x = a\sqrt{1 - z^2/C^2} \, \cos^{-1}(z/\lambda C)$$

since the curve is symmetric with respect to the z-axis. The maximum x value, x_m, will occur at the point where

$$\left. \frac{dx}{dz} \right|_{\substack{x=x_m \\ z=z_m}} = 0 = \left[-\frac{az}{C^2} \frac{\cos^{-1}(z/\lambda C)}{\sqrt{1 - z^2/C^2}} - \frac{a\sqrt{1 - z^2/C^2}}{\sqrt{\lambda^2 C^2 - z^2}} \right]_{\substack{x=x_m \\ z=z_m}} \tag{2.33}$$

where z_m is the z-coordinate of the maximum point.

Equation (2.33) can be written as

$$\left[\frac{z}{C^2} \cdot \frac{a\sqrt{1 - z^2/C^2}\ \cos^{-1}(z/\lambda C)}{\sqrt{1 - z^2/C^2}} + \frac{a}{C} \cdot \frac{\sqrt{1 - z^2/C^2}}{\sqrt{\lambda^2 - z^2/C^2}} \right]_{x_m, z_m} = 0$$

or

$$\left[\frac{z_m}{C^2} \cdot \frac{a\sqrt{1 - z_m^2/C^2}\ \cos^{-1}(z_m/\lambda C)}{\sqrt{1 - z_m^2/C^2}} + \frac{a}{C} \cdot \frac{\sqrt{1 - z_m^2/C^2}}{\sqrt{\lambda^2 - z_m^2/C^2}} \right]_{x_m} = 0 \qquad (2.34)$$

But

$$a\sqrt{1 - z_m^2/C^2}\ \cos^{-1}(z_m/\lambda C) = x_m \qquad (2.35)$$

$$1 - z_m^2/C^2 = x_g^2/a^2 \qquad (2.36)$$

where x_g is the x-coordinate of the generating ellipse when $z = z_m$. Putting (2.35) and (2.36) into (2.34), we have

$$\frac{x_m z_m a^2}{C^2 x_g^2} = -\frac{1}{C} \cdot \frac{x_g}{\sqrt{\lambda^2 - z_m^2/C^2}}$$

$$\lambda^2 - \frac{z_m^2}{C^2} = \left(\frac{C x_g^3}{x_m z_m a^2} \right)^2$$

Thus

$$\lambda = \left[\left(\frac{C x_g^3}{x_m z_m a^2} \right)^2 + \left(\frac{z_m}{C} \right)^2 \right]^{1/2} \qquad (2.37)$$

As in case 1, the value of λ determined from (2.37) may be subject to errors and deviations and may be smaller than 1 occasionally. In this case we have to set $\lambda = 1$. It is also possible to construct a numerical routine to determine a, C, and λ by digital computer, as will be demonstrated in Section 2.6.4.

Sometimes it is desirable to calculate x_m and z_m from a given combination of a, C, and λ. This is given in item iv, Section 2.6.4.

2.6.3. Examples and Discussion

Using the above steps, I have tried to fit many conical graupel and hailstones documented in the literature. Figure 2.20 shows two examples taken from Mason (1971) and one from Iribarne and Cho (1979). It appears that the fittings by Eq. (2.25) are reasonably close to the actual shapes, and that the most common values of λ lie between 1 and 2. Figure 2.21 shows an example of fitting for a falling large raindrop in an environment with and without a vertical electric field. The photographs are taken from Pruppacher et al. (1982).

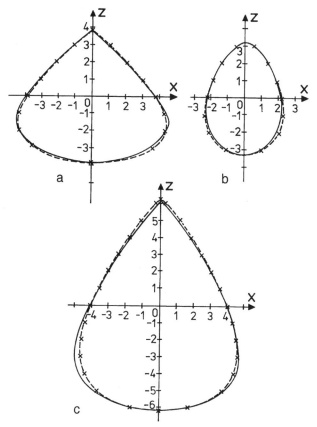

FIG. 2.20. Examples of fitting conical graupel and hailstones by Eq. (2.25). (a) $a = 2.42$, $c = 3.80$, and $\lambda = 1.0$. (b) $a = 1.40$, $c = 3.20$, and $\lambda = 1.72$. (c) $a = 2.55$, $c = 6.20$, and $\lambda = 1.045$. Original particle photographs simulated in (a) and (b) are taken from Mason (1971, Figs. 6.20 and 6.21) whereas (c) is taken from Iribarne and Cho (1980, Fig. V-21).

Note that Eq. (2.25) can also be used to describe spheres and spheroids, provided we replace $\cos^{-1}(z/\lambda C)$ by unity. The remaining equation then describes an ellipse. By rotating the ellipse about the z-axis, we obtain spheroids. If $a < C$, prolate spheroids result. If $a > C$, oblate spheroids result. In the special case where $a = C$, spheres result.

While Eq. (2.25) appears to be a reasonably good approximation to many conical hydrometeors, one also may note that it still represents an idealization. Natural conical hydrometeors, especially conical graupel and hailstones, may differ considerably from the idealized conical shape. They may not possess rotational symmetry and may have a rough surface. However, in many simple theoretical

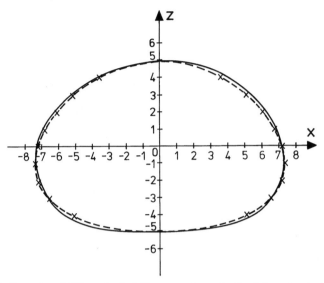

FIG. 2.21. Examples of fitting a freely falling water drop of radius 2.3 mm by Eq. (2.25) with $a = 4.58, c = 5.0$, and $\lambda = 14.79$. Original drop photograph simulated is taken from Pruppacher *et al.* (1982).

studies, such complications can be disregarded, at least to a first order of approximation. Equation (2.25) also provides a convenient way to categorize the dimensions of observed conical hydrometeors.

2.6.4. Cross-Sectional Area, Volume, and Area of the Surface of Revolution

The cross-sectional area, volume, and surface area are among the most important quantities that determine the physical properties of hydrometeors such as their liquid/solid water contents, heat and vapor diffusion rates, and so forth. We can use Eq. (2.25) to estimate the above geometric quantities for conical hydrometeors. One may refer to Courant and John (1965) for the formula used below.

(i) Cross-Sectional Area

We restrict ourselves here to the calculation of axial cross-sectional area of a conical hydrometeor. This area is

$$A = 2 \int_{-c}^{c} x \, dz = 2a \int_{-c}^{c} \sqrt{1 - \frac{z^2}{C^2}} \, \cos^{-1}\left(\frac{z}{\lambda C}\right) dz \qquad (2.38)$$

It turns out that this area is independent of λ, with

$$A = \pi^2 aC/2 \tag{2.39}$$

The derivation of (2.39) is given in Appendix A.

(ii) Volume

While the axial cross-sectional area is independent of λ, the volume does depend on λ. The volume is given by

$$V = \int_{-c}^{c} \pi x^2 \, dz = \pi a^2 \int_{-c}^{c} \left(1 - \frac{z^2}{C^2}\right) \left[\cos^{-1}\left(\frac{z}{\lambda C}\right)\right]^2 dz \tag{2.40}$$

There are two approaches in this integration: a series representation or an exact solution. The former has the benefit of faster calculation with relatively small error; but the latter is more elegant in the expression. The details of the calculation are given in Appendix B.

(A) Series Representation

The results are:

(a) When $\lambda = 1$

$$V = 3.6167 \pi a^2 C \tag{2.41}$$

(b) When $\lambda = \infty$,

$$V = 3.2889 \pi a^2 C \tag{2.42}$$

(c) When $1 < \lambda < \infty$,

$$V \approx \pi^2 aC \left(\beta_0 + \frac{\beta_2}{\lambda^2} + \frac{\beta_4}{\lambda^4} + \frac{\beta_6}{\lambda^6} + \frac{\beta_8}{\lambda^8} + \frac{\beta_{10}}{\lambda^{10}}\right) \tag{2.43}$$

where

$$\begin{aligned}
\beta_0 &= 3.2889 \\
\beta_2 &= 0.2667 \\
\beta_4 &= 0.0382 \\
\beta_6 &= 0.0113 \\
\beta_8 &= 0.0046 \\
\beta_{10} &= 0.0024
\end{aligned} \tag{2.44}$$

Since $\lambda \geq 1$, the largest error that can occur in using Eq. (2.43) is when $\lambda = 1$. But even in this case, the volume calculated from Eq. (2.43) is

$$V_{\lambda=1} = 3.6121\pi a^2 C$$

which deviates from Eq. (2.41) by just 0.13%. In view of the approximate nature of using Eq. (2.25) to describe the shape of a conical hydrometeor, this error is completely negligible.

(B) Exact Solution

The exact solution to (2.40), obtained by Magradze and Wang (1995), is

$$V = \pi a^2 C \left\{ \frac{\pi^2}{3} + \frac{4}{3} \left[\sin^{-1} \left(\frac{1}{\lambda} \right) \right]^2 - \left(\frac{8\lambda^2 - 32}{9} \right) \sqrt{\lambda^2 - 1} \sin^{-1} \left(\frac{1}{\lambda} \right) \right.$$
$$\left. + \left(\frac{24\lambda^2 - 104}{27} \right) \right\}$$

(2.45)

The details of the derivation are given in Appendix C.

(iii) Area of the Surface Revolution

The area of the surface of revolution can be calculated from Guldin's rule:

$$A_r = 2\pi \int_{-c}^{c} x \sqrt{1 + \left(\frac{dx}{dz} \right)^2} \, dz$$

(2.46)

This equation can be transformed into

$$A_r = -2\pi a C \lambda \int_{\cos^{-1}(-1/\lambda)}^{\cos^{-1}(1/\lambda)} v \left\{ f(v) \sin^2 v + \frac{a^2}{\lambda^2 C^2} \left[\frac{\lambda^2}{2} v \sin 2v + f(v) \right]^2 \right\}^{1/2} dv$$
$$= -2\pi a C \lambda I_r$$

(2.47)

where

$$v = \cos^{-1} \left(\frac{z}{\lambda C} \right)$$

(2.48)

and

$$f(v) = 1 - \lambda^2 \cos^2 v$$

(2.49)

The integral I_r in Eq. (2.47) was evaluated out numerically, and the result is surprisingly simple. For λ up to as large as 10^5, I_r is almost exactly equal to

$(-\pi/\lambda)$. Therefore for the regular range of λ (usually close to 1), the area of the surface of revolution is almost exactly

$$A_r = 2\pi^2 aC \tag{2.50}$$

to a high degree of accuracy. However, numerical calculation does show that I_r decreases slowly with increasing λ. As λ becomes infinity, the area A_r becomes

$$A_{r,\infty} = \pi^2 a \left[\frac{\pi a}{2} + \frac{\sin^{-1} KC}{K} \right] \tag{2.51}$$

where

$$K^2 = \left(\frac{1}{C^2} - \frac{\pi^2 a^2}{4C^4} \right) \tag{2.52}$$

Equation (2.51) can be easily checked by letting $\lambda \to \infty$ with $C = \pi a/2$. Then the second term in the brackets vanishes and we have a sphere with surface area $4\pi C^2 = \pi^3 a^2 = 2\pi^3 aC$. $A_{r,\infty}$ is smaller than $2\pi^2 aC$. For example, if we let $a = 5$ mm and $C = 10$ mm, then $A_{r,\infty} = 919.7$ mm^2 whereas $2\pi^2 aC = 986.96$ mm^2, a difference of 6.8%. Again, in view of the fact that these are all approximations, we can probably live with this error.

(iv) Cross-Sectional Area Normal to the Direction of Fall

The cross-sectional area normal to the direction of fall is one of the important dynamic quantities of conical particles. If the conical particle falls in the direction of its axis of symmetry, this area is simply

$$\pi X_m^2$$

where X_m was defined in Section 2.6.2. The formulation in Section 2.6.2 is for determining a, C, and λ from a photograph of a particle. Here we shall investigate a converse problem, namely, determining X_m and z_m for a given a, C, and λ. The most straightforward method would be to put Eqs. (2.33) and (2.25) in the computer directly and solve for X_m and z_m by an iterative method. In the following, we provide an alternative method.

We first note that the value of z_m (and thus the corresponding X_m) must be such that Eq. (2.33) holds. By letting $\cos u_m = (z_m/\lambda C)$ we can transform Eq. (2.33) into the following equation:

$$u_m \cos u_m \sin u_m - \cos^2 u_m = -\frac{1}{\lambda^2} \qquad (0 < u_m < \pi) \tag{2.53}$$

while at the same time one must remember that $\lambda^2 \cos^2 u_m \leq 1$.

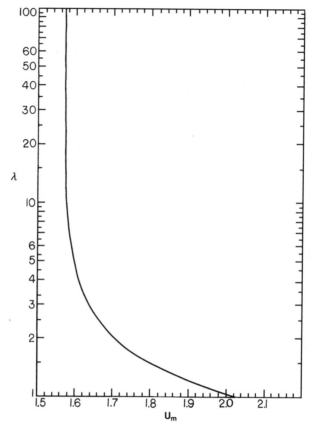

FIG. 2.22. Values of λ and the corresponding u_m from Eq. (2.54).

The possible solutions are in the range between $\pi/2$ (when $\lambda \to \infty$) and 2.028758 (when $\lambda = 1$). Figure 2.22 gives a curve of λ versus u_m. Thus for a given λ we can find a corresponding value of u. The value of z_m is then given by

$$z_m = \lambda C \cos u_m \qquad (2.54)$$

Putting this z_m, along with a, C, and λ, into Eq. (2.25), we obtain X_m. As an example, the conical graupel in Figure 2.20b has $a = 1.40$, $C = 3.20$, and $\lambda = 1.72$. From Figure 2.22, the value of u_m corresponding to $\lambda = 1.72$ is 1.75. Thus

$$z_m = (1.72)(3.20)\cos(1.75) = -0.9811$$

and from Eq. (2.25) the corresponding X_m is

$$X_m = (1.40)[1 - (-0.9811/3.20)^2]^{1/2}(1.75) = 2.3320$$

X_m is related to the characteristic lengths used by several previous investigators in describing conical particles. For example, Jayaweera (1971) used a characteristic length L^* defined by Pasternak and Gauvin (1960); L^* is the ratio of the total surface area divided by its perimeter P normal to the flow. In terms of a, C and λ, this would be equivalent to

$$L^* = \frac{A_r}{2\pi X_m} = \frac{2\pi^2 aC}{2\pi X_m} = \frac{\pi aC}{X_m} \qquad (2.55)$$

Other investigators, such as List and Schemenauer (1971) and Heymsfield (1978), used a characteristic length which is essentially $2X_m$.

Using a technique based on Eq. (2.25), Wang et al. (1987) analyzed the shape and size distributions of an ensemble of 679 hailstones collected by Dr. Nancy Knight of the National Center for Atmospheric Research (NCAR) during a 22 June 1976 hailstorm at Grover, Colorado. They demonstrated that the resulting distributions have relatively simple statistical behavior and thus the technique is useful for studying geometric properties of conical particles.

2.6.5. Three-Dimensional Expression of Conical Particles with Circular and Elliptical Cross Sections

In the discussions of the previous few sections, we used Eq. (2.25) to represent the axial cross sections of conical particles. This simplicity assumes that these horizontal cross sections of are circles. It has been pointed out by some investigators (Charles Knight, private communication; Albert Waldvogel, private communication) that many hailstones have elliptical horizontal cross sections. Hence it is desirable to have a formula describing such conical particles.

This can be easily done by first writing down the 3-D form of Eq. (2.25) and then generalizing it to include the case of elliptical cross sections. The 3-D form of Eq. (2.25) is simply

$$\frac{x^2}{[a\cos^{-1}(z/\lambda C)]^2} + \frac{y^2}{[a\cos^{-1}(z/\lambda C)]^2} + \frac{z^2}{C^2} = 1 \qquad (2.56)$$

An example of the 3-D body represented by (2.56) is shown in Fig. 2.23. The horizontal cross section of a conical body at a given z, specified by (2.56), is a circle of radius

$$a[\cos^{-1}(z/\lambda C)]$$

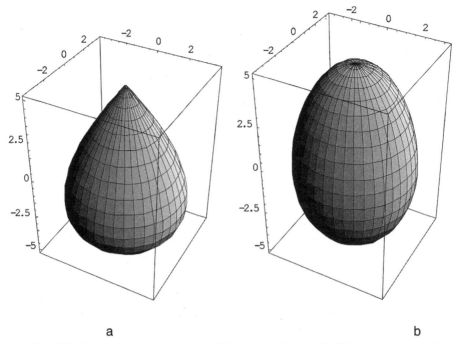

FIG. 2.23. Two z-axisymmetric conical particles generated by Eq. (2.64) by setting $a = c = 1$. (a) $\lambda = 1$. (b) $\lambda = 5$.

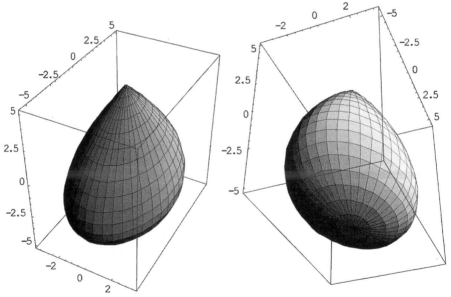

FIG. 2.24. A conical body with elliptical x–y cross sections generated by Eq. (2.62) with $a = 2$, $b = 1$, and $c = 5$.

It is now easy to generalize (2.56) to a conical body with elliptical horizontal cross sections. This is

$$\frac{x^2}{[a\cos^{-1}(z/\lambda C)]^2} + \frac{y^2}{[b\cos^{-1}(z/\lambda C)]^2} + \frac{z^2}{C^2} = 1 \qquad (2.57)$$

This form makes it clear that the horizontal cross section of a conical body specified by (2.57) at a given z is an ellipse of semi-axes $[a\cos^{-1}(z/\lambda C)]$ and $[b\cos^{-1}(z/\lambda C)]$ in the x and y directions, respectively. Figure 2.24 shows an example of such a particle.

The conical body as given by Eq. (2.56) has been used for the calculations of scattering by microwaves at 35.8 GHz. For details, see Sturnilo et $al.$ (1995).

3. Hydrodynamics of Small Ice Particles

3.1. Fall Attitude of Ice Particles

Ice particles are never truly stationary in clouds—they are always falling relative to the vertical air current. Since air is a viscous medium, the motion of ice crystals in it will produce complicated flow fields around the crystals. These flow fields, in turn, influence the way the ice crystals are falling.

When ice crystals are small, they fall in a steady manner. Observations show that small ice crystal plates or other planar crystals fall with their basal planes oriented horizontally, while small ice columns fall with their long axes horizontal. The flow fields around the falling crystals are said to be steady in this situation because they are independent of time.

As ice crystals grow larger, their fall attitude may become unsteady. Eddy shedding may start to occur in the downstream region. Eventually, they may fall in fairly complicated fashion; for example, large plates may fall in a zigzag manner, as anyone who has been through a snowstorm would never forget. Ice columns may perform rotation, frittings, and other unsteady fall motion (Jayaweera and Mason, 1965). For a more detailed description of these motions, the reader is referred to Pruppacher and Klett (1997).

Any quantitative investigation of the growth and dissipation of these ice particles will have to take their motion into account. For example, due to the existence of flow fields, the collision efficiencies between ice crystals and supercooled droplets (the riming process) can be significantly different from that deduced by just considering the geometry of the particles alone. Similarly, the diffusional growth of ice crystals will be influenced by their motion through the so-called ventilation effect (Pruppacher and Klett, 1997, Chap. 13).

In principle, the flow fields can be obtained either by experimental measurements or theoretical calculations. Experimental methods are preferred if they can be done properly. The reality is, however, that it is very difficult to measure the flow fields for a wide variety of cloud and precipitation particles, which vary greatly in size and shape, and for a wide range of atmospheric conditions. Since ice crystals are nonspherical, the measurements are especially difficult to perform. The other option is to perform theoretical calculations to see if the calculations can be done efficiently based on a realistic model. The advent of fast computers makes this option a viable choice, and the calculations can often be done at relatively economic cost in comparison with experimental measurements.

The theme of this section is the theoretical computations of flow fields around falling ice crystals. These flow fields can be either steady or unsteady depending on their Reynolds numbers (Re $= u_\infty \, d/\nu$). Calculations are done using realistic ice crystal models. The results compare favorably with the few experimental measurements available, demonstrating that theoretical computation is indeed a viable means of determining the flow fields.

3.2. Review of Previous Studies

Most of the previous theoretical studies of hydrodynamics relevant to cloud and precipitation particles have been reviewed and summarized by Pruppacher and Klett (1978) and Clift et al. (1978). Among the earlier studies relevant to this area are the analytical and semiempirical study of Stokes (see, for example, Happel and Brenner, 1965; Yih, 1969), Oseen (1910), Goldstein (1929), and Carrier (1953) regarding flow past rigid spheres, and Hadamard (1911) and Rybczinski (1911) regarding fluid spheres. Refinements of these early studies were made by many fluid dynamicists. However, it was soon realized that these analytical solutions can be applied only to a limited range of real atmospheric conditions, and producing results useful to cloud physics necessitates prescribing initial and/or boundary conditions that are more complicated and closer to realistic cloud environments. These problems would be very difficult to solve analytically and are indeed most conveniently solved by numerical methods. Thus, Jenson (1959), Hamielec et al. (1967), Le Clair et al. (1970), and Pruppacher et al. (1970) started to perform numerical calculations of fields for incompressible flow past rigid and liquid spheres.

Spherical problems, especially at low Reynolds numbers, are largely relevant only to cloud drops. Most other cloud and precipitation particles are prominently nonspherical. Large raindrops have relatively flat bottoms and round tops, like hamburger buns. Columnar ice crystals, dendrites, and conical graupel are certainly far from spherical. There is clearly a need to determine the flow fields around nonspherical hydrometeors. Some investigators have carried out a few such cases.

Again, more realistic solutions were obtained by numerical methods. For instance, the flow past infinitely long cylinders, which are often used to approximate flow fields around ice columns, were obtained by numerous researchers (e.g., Thom, 1933; Dennis and Chang, 1969, 1970; Hamielec and Raal, 1969; Takami and Keller, 1969; Schlamp et al., 1975). The numerical flow fields around thin oblate spheroids, used to approximate hexagonal ice plates, were obtained by Rimon and Lugt (1969), Masliyah and Epstein (1971), and Pitter et al. (1974).

All the studies mentioned above have two things in common. First, they all treated steady-state flow fields, which are only applicable to the motion of cloud and precipitation particles at low Reynolds numbers. Second, they treated only two-dimensional problems. In that Reynolds number range, the particles fall steadily, and therefore the flow fields around them are also independent of time. However, when these particles grow larger, they start to show unsteady fall behavior and create unsteady flow fields characterized first by the shedding of eddies downstream and then by turbulent eddies when the Reynolds numbers become sufficiently large. Undoubtedly, if we are to understand the unsteady motion of these particles and their effect on cloud growth, we need to determine these unsteady flow fields. This amounts to solving the unsteady Navier–Stokes equations with appropriate initial and boundary conditions. In addition, the flow fields around most real ice crystals are actually three-dimensional in nature even when the flow is steady. For example, the steady flow past a hexagonal plate does not really possess azimuthal symmetry, as would be the case for a circular disk or a thin oblate spheroid. The flow past a cylinder of finite length is even more asymmetric owing to the presence of a cylindrical surface and two plane end surfaces. When the flow becomes unsteady, of course, the asymmetry becomes even more pronounced.

In the following sections, I present some numerical techniques and the resulting flow fields around a few types of nonspherical ice particles in the low-to-medium (from 0.1 to about 200) Reynolds number range. These are taken from Ji and Wang (1989, 1991) and Wang and Ji (1997).

3.3. The Physics and Mathematics of Unsteady Flow Fields around Nonspherical Ice Particles

3.3.1. Streamfunction versus Momentum Equation Formulation

In this section, we discuss the conceptual setup of the problems for unsteady flow past nonspherical ice crystals and the numerical schemes that we used to solve them. In the treatment of two-dimensional steady-state incompressible flow problems, it is common to formulate the problems in terms of a scalar streamfunction ψ. The benefit of doing so is that only a single dependent scalar variable needs to be solved for, and the components (e.g., u, v) of the flow velocity vector **V** can be derived from ψ. On the other hand, using the original momentum equation formulation would require solving for two dependent variables.

Unfortunately, the attractiveness of the streamfunction formulation disappears for three-dimensional flows. While it is still possible to define a streamfunction, this function will be a vector instead of a scalar (see, for example, Anderson *et al.*, 1984). This means that three separate component equations of the streamfunction need to be solved instead of one. Thus there is no advantage of the streamfunction formulation over the original momentum equations. In the present study, the momentum equation formulation is used.

3.3.2. The Incompressible Navier–Stokes Equations and the Initial and Boundary Conditions

We shall treat three relatively simple ice crystal shapes: the columnar ice crystals (approximated by finite circular cylinders), hexagonal ice plates, and broad-branch crystals. Figure 3.1 shows a schematic sketch of these three types of crystals. The quantity a represents the "radius" of the ice crystals as defined in the figure. We shall also assume that these ice crystals fall with their broad dimensions oriented in the horizontal direction, which is known to be the common fall orientation of many medium-sized ice crystals (Pruppacher and Klett, 1997). It is known that large ice crystals also exhibit zigzag fall attitudes, but that is not simulated here owing to limited computer resources. The schematic configuration of the theoretical problem considered here is shown in Figure 3.2.

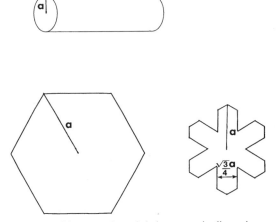

FIG. 3.1. The three types of ice crystals and their geometric dimensions considered in this study.

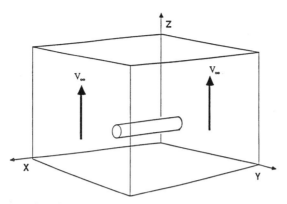

FIG. 3.2. The general configuration of the theoretical problem. Ice crystals are assumed to fall with their long axes in the horizontal direction.

To facilitate the numerical analysis, we first introduce the following dimensionless variables:

$$x' = \frac{x}{a}, \quad \mathbf{V}' = \frac{\mathbf{V}}{|\mathbf{V}_\infty|}, \quad t' = \frac{t|\mathbf{V}_\infty|}{a}, \quad P' = \frac{P}{\rho|\mathbf{V}_\infty|^2}, \quad \mathrm{Re} = \frac{2|\mathbf{V}_\infty|a}{\nu}$$

$$(3.1)$$

where x (or y, z) is one of three Cartesian coordinates, $\mathbf{V}$ is the fluid velocity, $\mathbf{V}_\infty$ is the freestream velocity (which is equal to the terminal fall velocity of the ice crystal), P is the dynamic pressure, ν is the kinematic viscosity of the fluid, ρ the air density, and Re is the Reynolds number relevant to the flow. All primed quantities are nondimensional. Using these dimensionless variables, we can write down the nondimensional Navier–Stokes equation and continuity equation as (after dropping the primes)

$$\frac{\partial \mathbf{V}}{\partial t} + \mathbf{V} \cdot \nabla \mathbf{V} = -\nabla P + \frac{2}{\mathrm{Re}} \nabla^2 \mathbf{V} \tag{3.2}$$

$$\nabla \cdot \mathbf{V} = 0 \tag{3.3}$$

The ideal boundary conditions appropriate for the present problems are

$$\mathbf{V} = 0 \quad \text{at the surface of the ice crystal} \tag{3.4}$$

and

$$\mathbf{V} = \mathbf{e}_z \quad \text{at infinity} \tag{3.5}$$

where $\mathbf{e}_z$ is the unit vector in the general flow direction. In real numerical computations, of course, the domain is always finite and condition (3.5) can only be taken to mean that the velocity is constant at an outer boundary that is sufficiently far away from the crystal. It is difficult at present to predetermine on purely theoretical

TABLE 3.1 OUTER BOUNDARIES OF THE COMPUTATIONAL DOMAINS FOR THE
THREE CRYSTAL CASES[a]

Boundaries	Columnar crystal	Hexagonal plate	Broad-branch crystal
Upstream	12.7	3.3	1.7
Lateral			
Lengthwise	25.5	6.74	10.10
End-on	23.5	6.86	11.09
Downstream	63.0	14.3	15.27

[a] The radius is 1.

ground how far the distance should be in order to be called "sufficiently far." We did this by trial and error, and considered the outer boundary far enough away as long as the computed results do not change by more than a few percent as we move the boundary farther out. Similar treatment was done for all outer boundaries. Table 3.1 shows the locations of the upstream, downstream, and lateral boundaries for determining the numerical flow fields in the three cases.

While condition (3.5) is approximately valid at the upstream and lateral boundaries, it is usually not valid at the downstream boundaries. This is because, at the relatively high Reynolds number range investigated here, the shedding of eddies may occur. The disturbances often propagate downstream for a long distance. Thus, at the downstream boundary, (3.5) is replaced by a weaker condition

$$\frac{\partial \mathbf{V}}{\partial z} = 0$$

The pressure field can be determined from the Navier–Stokes equation at all boundaries except at the downstream boundary, where the condition

$$\frac{\partial P}{\partial z} = 0$$

is used. Since we are dealing with unsteady flow here, we also need initial conditions to close the equations. The initial conditions (at $t = 0$) are $P = 0$ and $\mathbf{V} = \mathbf{e}_z$ everywhere except at the surface of the crystal, where we require $\mathbf{V} = 0$ (nonslip condition).

3.3.3. Generation of Unsteady Flow Features

Although the Navier–Stokes equation (3.2) is written as a time-dependent equation, this does not mean that the computational results will always result in time-dependent flow features such as the shedding of eddies. Indeed, Dennis and Chang (1970) have shown that for flow past two-dimensional cylinders starting with symmetric initial conditions, the eddy shedding does not occur even at Reynolds numbers as high as 1000. In order to generate these time-dependent, or unsteady, features, it is necessary to implement an asymmetric initial perturbation field.

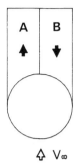

FIG. 3.3. The initial perturbation imposed on the steady flow field in order to generate time-dependent flow behavior. The magnitude of perturbation in regions A and B is 30% of the free-stream velocity but the directions of the perturbations are opposite.

There are many ways of implementing this perturbation. For example, Braza *et al.* (1986) achieved this on a two-dimensional cylinder by performing a rotation of the cylinder along its axis. In the present study, we achieve this by implementing a velocity perturbation of magnitude $0.3V_\infty$ in the downstream region immediately behind the crystal to the steady-state solutions, as shown in Figure 3.3. The directions of the perturbation are opposite each other in the regions A and B so as to form a shear along the central plane of the flow. As we shall see, at high enough Reynolds numbers, this perturbation will generate a periodic eddy shedding pattern in the simulated flow. On the other hand, the perturbation will be damped out in a short time if the Reynolds number is low.

3.4. The Numerical Scheme

To solve Eqs. (3.2) and (3.3) with the appropriate initial and boundary conditions, we adopt a numerical approach utilizing the finite difference method. It is also necessary to set up a mesh grid. Due to the more complicated shapes of these ice crystals, it is decided that the simplest way to set up grids is to use the Cartesian coordinate system. In order to prescribe the inner boundary conditions with adequate precision, the grid spacing near the crystal surface has to be small. On the other hand, the grid spacing far from the crystal can be larger to save computing time. This results in nonuniform grids used in the present study, as shown in Figure 3.4.

As indicated before, the primitive velocity formulation of the Navier–Stokes equation is adopted for this study. The velocity at each time step is obtained by a predictor–corrector method. First, the velocity predictor $\mathbf{V}^*$ is determined by solving the following equation:

$$\frac{\mathbf{V}^* - \mathbf{V}}{\Delta t} + (\mathbf{V}^n \cdot \nabla)\mathbf{V}^n = \frac{2}{\mathrm{Re}}\nabla^2\mathbf{V}^n \tag{3.6}$$

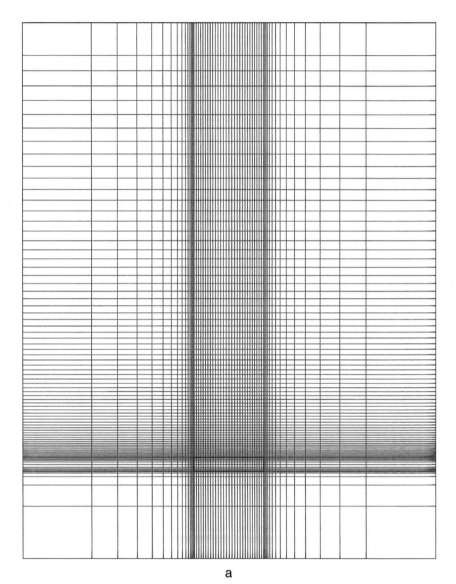

a

FIG. 3.4. The nonuniform grids used for numerically solving the Navier–Stokes equations for flow past a columnar ice crystal: (a) broadside view; (b) end view.

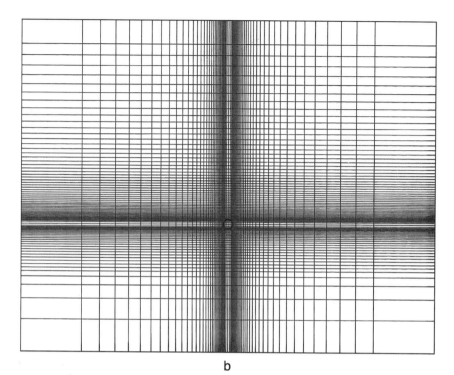

b

FIG. 3.4. *Continued.*

where $\mathbf{V}^n$ is the velocity solved at time step n and Δt is the time increment. The pressure at time step $n + 1$ is then given by

$$\nabla^2 P^{n+1} = \frac{\nabla \cdot \mathbf{V}^*}{\Delta t} \tag{3.7}$$

(see, for example, Peyret and Taylor, 1983). Finally the velocity at time step $n + 1$ is determined by

$$\frac{\mathbf{V}^{n+1} - \mathbf{V}^*}{\Delta t} = \nabla P^{n+1} \tag{3.8}$$

A scheme of quadratic upstream interpolation for convective kinematics (QUICK) is used here (Leonard, 1979, 1983; Freitas *et al.*, 1985; Davis, 1984). Since a uniform grid system is obviously inefficient in the present problem (higher resolution is needed in regions near the cylinder), we adopted Freitas and colleagues' (1985) strategy of a modified QUICK version which applies to a nonuniform grid. The quadratic interpolation equation becomes one-dimensional if Leonard's CURVT terms are neglected. Thus, following Freitas *et al.*, let a function Φ of ξ be written as $\Phi = c_1 + c_2\xi + c_3\xi^2$, where ξ is the local coordinate of a control volume. The general quadratic interpolation equation is then integrated over each control volume face to generate the corresponding one-dimensional flux interpolation equations for a nonuniform grid. These equations are as follows:

For $u > 0$:

$$\Phi_e = \left\{ \frac{1}{2} + \frac{\Delta x_i / \Delta x_{i-1}}{4} \right\} \Phi_i - \left\{ \frac{\Delta x_i / (\Delta x_i + \Delta x_{i-1})}{4} \right\} \Phi_{i+1}$$

$$+ \left\{ \left[\frac{(\Delta x_i / 2 + \Delta x_i^2 / \Delta x_{i-1} / 4)}{(\Delta x_i + \Delta x_{i-1})} \right] - \frac{\Delta x_i / \Delta x_{i-1}}{2} \right\} \Phi_{i-1} + \frac{\Phi_{i+1}}{2} \quad (3.9)$$

$$\Phi_w = \left\{ \frac{1}{2} - \frac{\Delta x_{i-1} / (\Delta x_{i-2} + \Delta x_{i-1})}{4} \right\} \Phi_i - \left\{ \frac{1}{2} + \frac{\Delta x_{i-1} / \Delta x_{i-2}}{4} \right\} \Phi_{i-1}$$

$$+ \left\{ \left[\frac{(\Delta x_{i-1} / 2 + \Delta x_{i-1}^2 / \Delta x_{i-2} / 4)}{(\Delta x_{i-2} + \Delta x_{i-1})} \right] - \frac{\Delta x_{i-1} / \Delta x_{i-2}}{2} \right\} \Phi_{i-2} \quad (3.10)$$

For $u < 0$:

$$\Phi_e = \left\{ \frac{1}{2} - \frac{\Delta x_i / (\Delta x_{i+1} + \Delta x_i)}{4} \right\} \Phi_i + \left\{ \frac{1}{2} + \frac{\Delta x_i / \Delta x_{i+1}}{4} \right\} \Phi_{i+1}$$

$$+ \left\{ \left[\frac{(\Delta x_i / 2 + \Delta x_i^2 / \Delta_{i+1} / 4)}{(\Delta x_{i+1} + \Delta x_i)} \right] - \frac{\Delta x_i / \Delta x_{i+1}}{2} \right\} \Phi_{i-2} \quad (3.11)$$

$$\Phi_w = \left\{ \frac{1}{2} - \frac{\Delta x_{i-1} / \Delta x_i}{4} \right\} \Phi_i + \left\{ \left[\frac{(\Delta x_i / 2 + \Delta x_{i-1}^2 / \Delta x_i / 4)}{(\Delta x_i + \Delta x_{i-1})} \right] \right.$$

$$\left. - \frac{\Delta x_{i-1} / \Delta x_i}{2} \right\} \Phi_{i+1} - \left\{ \frac{\Delta x_{i-1} / (\Delta x_i + \Delta x_{i-1})}{4} \right\} \Phi_{i-1} + \frac{\Phi_{i-1}}{2} \quad (3.12)$$

where $\Delta x_i = (x_{i+1} - x_i)$.

Using the above operators, a component equation of Eq. (3.6) can be written as

$$\Phi_i^{n+1} = \Phi_i^n - \left[\frac{u_i / (\Delta x_{i-1} + \Delta x_i)}{2} \right] (\Phi_e - \Phi_w)$$

$$+ \frac{2[\Delta x_{i-1} \Phi_{i+1} + \Delta x_i \Phi_{i-1} - (\Delta x_{i-1} + \Delta x_i) \Phi_i]}{\Delta x_i / \Delta x_{i-1} / (\Delta x_i + \Delta x_{i-1})} \quad (3.13)$$

The three-dimensional QUICK scheme can be similarly constructed. The stability condition for one dimension is

$$\Delta t < \frac{4}{(\text{Re} \cdot \mathbf{V}_{\text{max}}^2)} \quad (3.14)$$

and

$$\frac{2\Delta t}{(\text{Re} \cdot \Delta x^2)} + \mathbf{V}_{\text{max}} \frac{\Delta t}{(4 \, \Delta x)} < 0.5 \quad (3.15)$$

where Δt is the time step and $\mathbf{V}_{\max}$ is the maximum velocity in the flow domain. To ensure stability in 3-D calculations, we took time steps, which are at least three times as small as those required by Eq. (3.14). Similarly, Δx's were chosen so that the right-hand side of Eq. (3.15) is three times smaller than required. As pointed put by Freitas *et al.* (1985), such a scheme results in second-order accuracy.

The Poisson equation for pressure, Eq. (3.7), is solved by the standard successive over relaxation (SOR) method described in Peyret and Taylor (1983) and Anderson *et al.* (1984).

The time step Δt used in the integration varies from 0.015 to 0.03 depending on the local grid spacing such that the stability criterion is satisfied. The smallest grid spacing was $\Delta x = 0.0775$. The largest grid size used was $59 \times 75 \times 89$. Naturally, a larger grid size will result in better accuracy but will increase the computing time considerably. The typical computing time for 10,000 time steps is on the order of a few hours on a Cray X/MP computer. The computation on a Cray-2 computer is somewhat faster. It appears that the SOR scheme in solving the pressure equation is the main bottleneck of the computation. The grid size used in this study represents a compromise between accuracy and available computing resource.

3.5. Results and Discussion

The size, aspect ratios, and Reynolds numbers of the columns, hexagonal plates, and broad-branch crystals are listed in Tables 3.2, 3.3, and 3.4, respectively. Their dimensions are chosen to overlap those adopted by some previous work (Schlamp *et al.,* 1975, 1976; Pitter *et al.,* 1973; Pitter and Pruppacher, 1974; Pitter, 1977; Miller and Wang, 1989) so that the results can be compared. In the following we discuss the results for each crystal type separately.

TABLE 3.2 DIMENSIONS OF COLUMNAR ICE CRYSTALS
TREATED IN THE PRESENT STUDY

| | Units (dimensionless) | | |
Re	Diameter (d)	Length (L)	L/d
0.2	2.0	2.85	1.43
0.5	2.0	2.85	1.43
0.7	2.0	3.08	1.54
1.0	2.0	3.33	1.67
2.0	2.0	4.44	2.22
5.0	2.0	6.67	3.33
10.0	2.0	10.00	5.00
20.0	2.0	16.67	8.33
40.0	2.0	12.58	6.29
70.0	2.0	25.32	12.66

PAO K. WANG

TABLE 3.3 DIMENSIONS OF HEXAGONAL ICE PLATES
TREATED IN THE PRESENT STUDY

	Units (dimensionless)		
Re	Diameter (d)	Thickness (h)	h/d
1.0	2.0	0.225	0.1125
2.0	2.0	0.177	0.0885
10.0	2.0	0.1265	0.06325
20.0	2.0	0.1034	0.0517
35.0	2.0	0.0863	0.04315
60.0	2.0	0.0725	0.03625
90.0	2.0	0.064	0.032
120.0	2.0	0.0576	0.0288

3.5.1. Comparison with Experimental Results

Before we present the complete results of the flow fields around falling ice columns, it is of interest to compare the numerically calculated and experimental measured results so that the validity of the numerical scheme can be checked to a certain degree. The main experimental results for this purpose come from Jayaweera and Mason (1965).

Previous experimental studies of both two- and three-dimensional flow past circular cylinders indicated that the flow remains steady up to Re $\approx$ 50 (e.g., Kovasznay, 1949; Jayaweera and Mason, 1965). It is therefore useful to perform calculations for flow past finite cylinders at Reynolds numbers below and above 50 to represent the steady and unsteady cases, respectively. We chose Re = 40 and 70 for this purpose. The aspect (diameter/length) ratio (d/l) is 0.159 for Re = 40 and 0.079 for Re = 70. These ratios are the same as that of the cylinders used in the experiment of Jayaweera and Mason (1965).

TABLE 3.4 DIMENSIONS OF BROAD-BRANCH CRYSTALS
TREATED IN THE PRESENT STUDY

	Units (dimensionless)		
Re	Diameter (d)	Thickness (h)	h/d
1.0	2.0	0.15	0.075
2.0	2.0	0.14	0.07
10.0	2.0	0.0914	0.0457
20.0	2.0	0.080	0.040
35.0	2.0	0.0667	0.033
60.0	2.0	0.060	0.03
90.0	2.0	0.052	0.026
120.0	2.0	0.047	0.0235

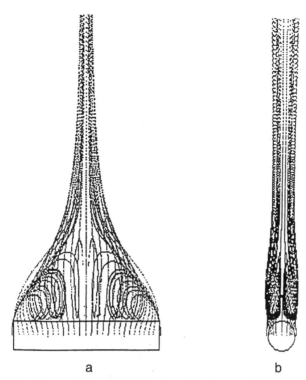

a   b

FIG. 3.5. Streak pattern, or "snapshot" field, of massless tracer particles for flow past an ice column for Re = 40: (a) broadside view; (b) end view.

In order to compare visually with the experimental photographs of Jayaweera and Mason (1965), positions of massless marker particles were calculated. These particles were originated from various places on the surface of the cylinder and new particles were introduced at each time step. Figure 3.5 shows the streak plots of these particles. Since the flow at this Reynolds number tends to converge to steady state (even if unsteady initially, as will be discussed later), the streaklines are essentially the same as streamlines. The pyramidal standing eddies shown in Figure 3.5a are similar to those described above. It is seen that the eddies consist of mainly fluid particles originated near the ends of the cylinder. This is due to the relatively large vorticities there that trap the particles. Particles originated in the center part of the cylinder follow relatively straight paths and quickly move to the downstream. Figure 3.5b shows the end-on view of particle streaks in the central yz plane. This view looks more like the familiar two-dimensional eddies attached to an infinitely long cylinder. Figures 3.5a and 3.5b show striking similarities to the photographs of tracer particles taken by Jayaweera and Mason (1965), as

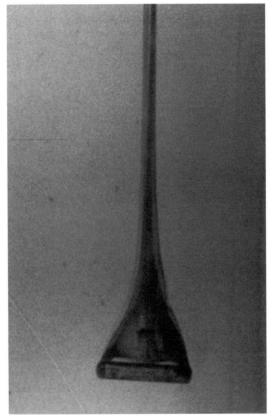

FIG. 3.6. Experimental photographs of a falling short cylinder at Re = 40. Broad-side view. (Photo courtesy of Dr. K. O. L. F. Jayaweera.)

shown in Figure 3.6. The characteristic pyramidal wake in the broadside view, the shape of the standing eddies, and the slightly diverging streak tails are well reproduced.

It is important to remember at this point that the trajectories and streaks of particles (whether massless or not) are the same in this case since the flow is steady. This is not true if the flow is unsteady.

Figure 3.5a also reveals an interesting feature of the flow in the pyramidal wake. It is obvious that tracer particles near the edges of the cylinder are subject to higher vorticity of the flow, hence tending to stay longer in the wake. Tracer particles near the center of the cylinder, on the other hand, would experience less vorticity and hence spend little time in the wake, instead, they go into the long tail region quickly. This phenomenon is illustrated even more clearly from Figure 3.7, where a top view of the tracer particle trajectories is shown. Here a tracer particle near the

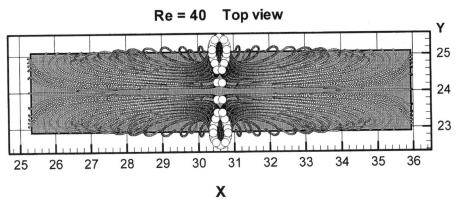

FIG. 3.7. Top view of the computed tracer particle trajectories around the falling cylinder for Re = 40. Larger particles represent those located at "higher" altitudes (i.e., more downstream). Particles originate at points one grid away from the cylinder surface.

center makes a simple curve and swiftly goes into the long tail region whereas a particle near the edge makes many loops before it reaches the long tail region.

As mentioned previously, experimental studies indicate that the flow past cylinders is steady at Re = 40. This would mean that small disturbances occurring in the field would be damped whereas at higher Reynolds numbers, the disturbances will tend to develop into unsteady flow patterns and result in shedding eddies. To see if this damping can be numerically simulated, we introduced an artificial disturbance in the flow as described in Section 3.3.3 and observed the development of this perturbed field.

Figure 3.8 shows the fluctuation of velocity in this perturbed field with time at a point in the wake. The perturbation was introduced at $t = 91$, indicated by the sharp downward spike. This came as a fluctuation in the velocity field, which becomes smaller and smaller as time goes on and is eventually damped out. Fluctuation of velocities at other points in the downstream were also traced and similar behavior was found. Figure 3.9 shows the computer marker particle streaks at $t = 120$. At this time, the initial disturbance has propagated a bit downstream and the field immediate to the cylinder is beginning to restore to the initial steady state. At $t = 165$, shown in Figure 3.10, the disturbance is nearly gone and the flow is almost completely restored to the original steady field. This demonstrates that for such a cylinder at Re = 40, the flow tends to become steady-state even under a perturbation as much as 30% of the freestream velocity. This also indicates that the steady flow field obtained in this calculation is a stable one. It will be of interest to see if there exists a critical perturbation magnitude that will set the flow into an unsteady pattern. This, however, is not yet done.

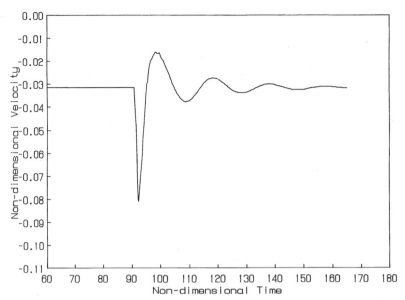

FIG. 3.8. Fluctuation of the nondimensional velocity with time at the point (38.0, 24.0, 18.9) in the wake of the flow field of the falling cylinder (Re = 40) considered here.

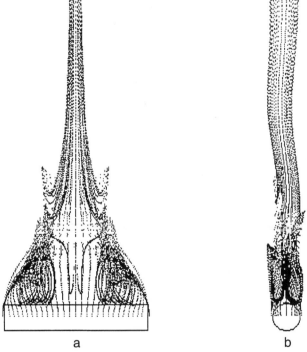

a   b

FIG. 3.9. Computed particle streaks for perturbed flow past the short cylinder at Re = 40 at t = 120. (a) Broadside view; (b) end view.

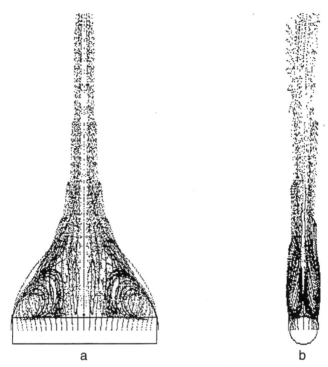

a   b

FIG. 3.10. Computed particle streaks for perturbed flow past the short cylinder at Re = 40 at $t = 165$. (a) Broadside view; (b) end view.

As mentioned earlier, at higher Reynolds numbers the flow past a finite cylinder becomes unsteady, as demonstrated by the experiment of Jayaweera and Mason (1965). Eddies begin to shed away from the cylinder and form the von Karmen vortex street. In order to study this eddy shedding, a perturbation of the same magnitude as above was again introduced into a steady flow field. For the purpose of a fair comparison with the Re = 70 case, we computed the flow past a cylinder exactly the same as before except for changing the Reynolds number to 70. Now the result is completely different from the previous case. Figure 3.11 shows a "snapshot" of the computed particle streaks at $t = 120$. Instead of the damping, the perturbation develops into a periodic oscillating pattern and shows no sign of damping in later times. Since the dimension of the cylinder is the same as before, the different result obtained here can only be interpreted as due to a different Reynolds number. This, together with the fact that other investigators have also successfully simulated the unsteady motion by different kinds of perturbations (e.g., Braza *et al.,* 1986), seems to confirm the notion that the periodic character of the flow is an intrinsic property of the Navier–Stokes equations and does not depend on the method of the perturbation.

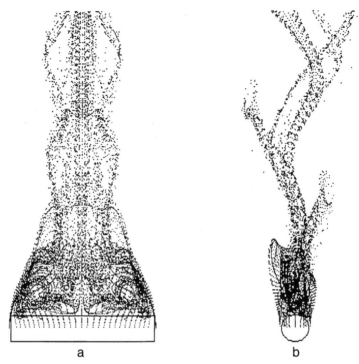

a   b

FIG. 3.11. Computed particle streaks for perturbed flow past the short cylinder at Re = 70 at $t = 120$. (a) Broadside view; (b) end view. The dimensions and aspect ratio of this cylinder are the same as that in Figures 3.9 and 3.10.

The result of the Re = 70 case described above cannot be compared directly with the experimental results of Jayaweera and Mason, however, because the aspect ratios of the two cases are different. In order to compare fairly with experimental results, we performed another set of calculations for Re = 70, but using a cylinder with an aspect ratio the same as that in Jayaweera and Mason (1965). The same perturbation was again introduced after a steady flow field was obtained. Again, the perturbation developed into periodic shedding of eddies. Figure 3.12 shows the fluctuation of velocity component v at a point in the wake. The abscissa represents the time starting with the perturbation. The periodic oscillation becomes a stable pattern at $t = 120$. The Strouhal number St ($= nd/V$, where n is the frequency of the oscillation) is about 0.138 which is close to but slightly below the value reported by Chilukuri (1987) for unsteady flow past a 2-D cylinder (see Fig. 3.13).

Figure 3.14 shows particle streaks for this case in the end and broadside views at $t = 300$. Both look strikingly similar to the experimental photographs taken by

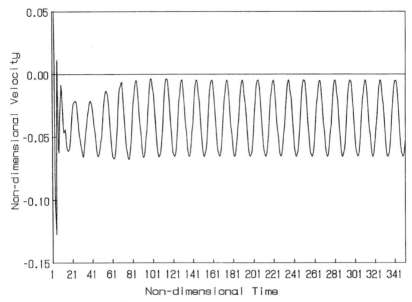

FIG. 3.12. Fluctuation of the nondimensional velocity with time at the point in the wake of the flow field of a falling cylinder (Re = 70). Note the aspect ratio of this cylinder is the same as that in Jayaweera and Mason (1965), but differs from that in Figures 3.9 and 3.10.

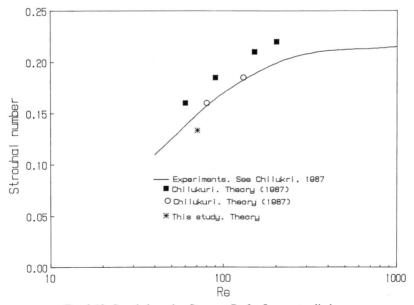

FIG. 3.13. Strouhal number St versus Re for flow past cylinders.

a   b

FIG. 3.14. Computed particle streaks for perturbed flow past the short cylinder at Re = 70 at
$t = 120$. (a) Broadside view; (b) end view. Dimensions of this cylinder are the same as that in
Figure 3.15.

Jayaweera and Mason (1965) for the same Reynolds number. Figure 3.15 shows an
experimental photograph for Re = 70 (K. L. O. F. Jayaweera, 1988, private communication). It is seen that both the computed angle extended by the vortex street
and positions of individual vortices agree excellently with experimental results.
Figure 3.16 shows the top view of the massless trace particle streaks (i.e., snapshot
of particle positions). Note that since the flow is unsteady, the particle streaks are
different from trajectories. An interesting contrast to the two-dimensional flow
case is illustrated in this figure. Here we see that particles originating near the
edges of the cylinder would first flow toward the center of the cylinders and go
upward at the same time. Because of the unsteady nature, the shedding of eddies
occurs in the downstream and hence the particle streaks also oscillate with their
eddy shedding. As the particles go up, their streaks would oscillate along an axis
perpendicular to the cylinder. In time, the spread of the tracer streaks will result
in a pattern perpendicular to the cylinder itself. This is a pure three-dimensional
feature. A true two-dimensional system, such as an infinitely long cylinder, would
result in a wake streak pattern still parallel to the cylinder itself.

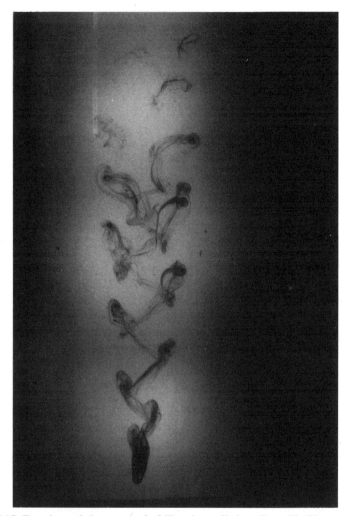

FIG. 3.15. Experimental photograph of a falling short cylinder at Re $= 70$. (Photo courtesy of Dr. K. O. L. F. Jayaweera.)

Figure 3.17 shows a few trajectories of tracer particles originated from different parts near the cylinder surface. Although some periodic features are also present, it is quite clear that the trajectories are quite different from the streaks due to the unsteady nature of the flow.

In short, the computational results presented in this section seem to agree with available experimental measurements to a good extent, suggesting that the numerical schemes used here are probably reasonable.

Re = 70   Top view

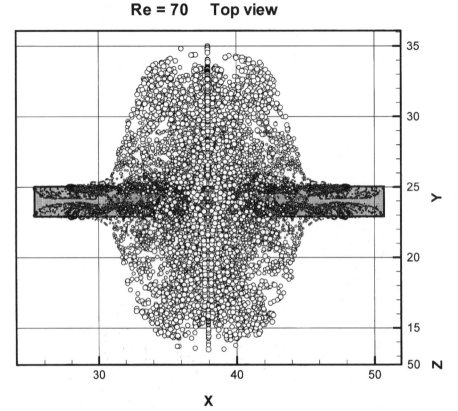

FIG. 3.16. Top view of the computed tracer particle trajectories for flow past a cylinder at Re = 70. Larger particles represent those located at "higher" altitudes (i.e., more downstream).

3.5.2. General Features of the Flow Fields around Falling Columnar Ice Crystals

We are now ready to present the computed results of the flow fields around falling columnar ice crystals. As mentioned earlier, this type of crystal is approximated by a circular cylinder of finite length. Because of the finite length, the cylinder, as well as the flow field around it, is no longer cylindrically symmetric. The dimensions and aspect ratios of the cylinders chosen for the computation are shown in Table 3.2 and are the same as those given by Schlamp *et al.* (1975) for Reynolds numbers between 0.2 and 20 and by Jayaweera and Mason (1965) for Reynolds numbers 40 and 70 (as seen in the last section). Several higher Reynolds numbers cases were also computed for the purpose of checking, but the details of these will not be discussed here. The aspect ratios of the cylinders specified by Schlamp *et al.*

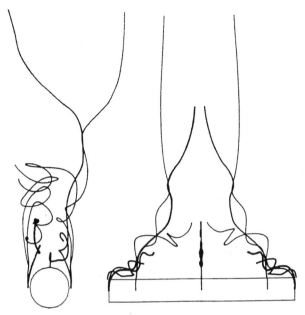

FIG. 3.17. Computed trajectories of a few selected tracer particles for flow past a cylinder at Re = 70.

(1975) are taken from the actual samples whose diameter-length relations were reported by Auer and Veal (1970). In all cases, the ice columns become longer as compared to the diameters as the Reynolds numbers increases.

As an example of the case of eddyless flow, Figures 3.18a–d show the velocity fields and the vorticity fields of flow past an ice column of Re = 2.0. The velocity fields in the central plane of the crystal in the broadside and end views, shown in Figures 3.18a and 3.18b, respectively, reveal a completely laminar flow pattern without any trace of an eddy. Note that the velocity vectors shown here are the projections of the three-dimensional velocities on the xz plane. The flow, of course, is not symmetric in the fore and aft direction, which is most evidently shown by the corresponding views of the vorticity fields in Figures 3.18c and 3.18d. Highest vorticity occurs in the two corners of the front edge.

The case of flow fields with standing eddies is amplified by that shown in Figures 3.19a–c for the flow past an ice column at Re = 40. As we have seen before, this is a steady flow case. In Figure 3.19a, the wake region looks triangular which, in three dimension, is actually pyramidal in shape, as we have seen in the tracer streaks before. The extent of the return flow region (i.e., the length of the eddy) is about 5.6 radii downstream, somewhat longer then the 4.8 radii for flow

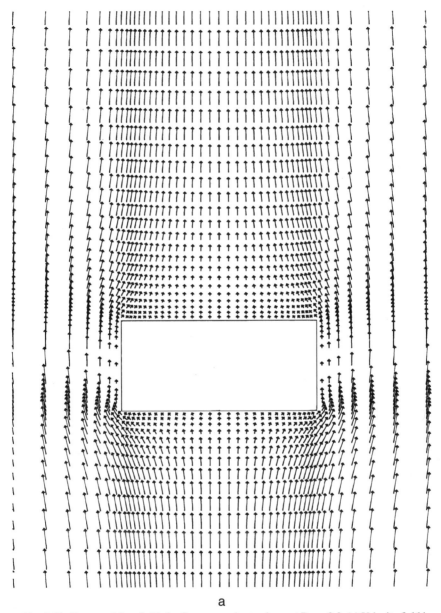

a

FIG. 3.18. Computed flow fields for flow past a short column at Re = 2.0. (a) Velocity field in the central cross section, broadside view. (b) Velocity field in the central cross section, end view. (c) Vorticity field in the central cross-section, broadside view. (d) Vorticity field in the central cross-section, end view.

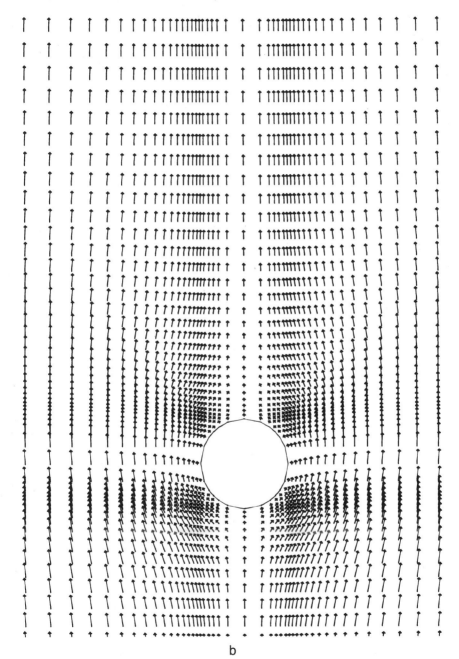

b

FIG. 3.18. *Continued.*

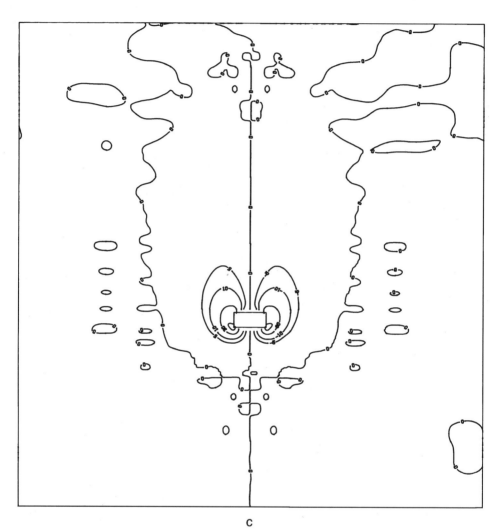

C

FIG. 3.18. *Continued.*

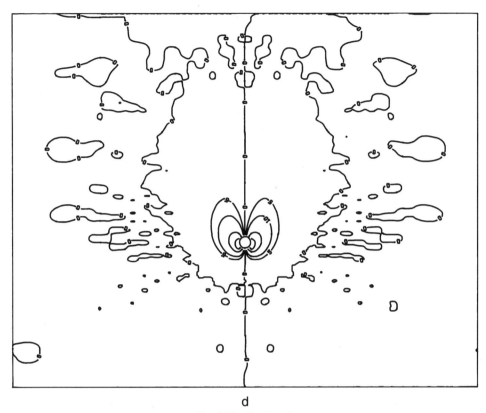

d

FIG. 3.18. *Continued.*

past an infinitely longer cylinder at the same Reynolds number (Pruppacher and Klett, 1978, p. 336).

Figures 3.19b and 3.19c show the end view of the velocity fields in the central and near-edge planes, respectively. The feature of the standing eddies is quite clear. It is also clear that the eddy size as well as the intensity in Figure 3.19c is much smaller than the eddy size in Figure 3.19b, consistent with the pyramidal structure model of the eddy.

The vorticity field in the central xz plane is shown in Figure 3.19d. Again, the maximum vorticities occur at the two front edges. Figure 3.19e shows the vorticity field in the central yz plane.

Figures 3.20a–d show the flow fields pertaining to the case of Re = 70 at $\lambda = 300$. As we have noted before, this is an unsteady flow case. Figure 3.20a shows the velocity field in the central xz plane. This periodic shedding of the eddies in the downstream region is clearly seen and has the same pattern as the tracer streaks shown in Figure 3.14a, as it should be. Figure 3.20b, on the other hand,

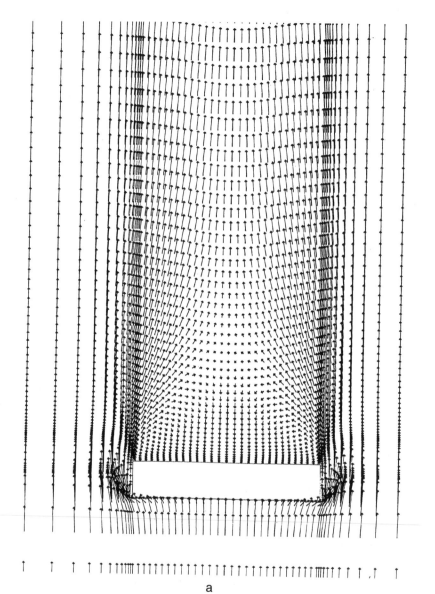

a

FIG. 3.19. Computed flow fields for flow past a short column at Re = 40. (a) Velocity field in the central cross section, broadside view. (b) Velocity field in the central cross section, end view. (c) Velocity field in a cross section near an end surface, end view. (d) Vorticity field in the central cross section, broadside view. (e) Vorticity field in the central cross section, end view.

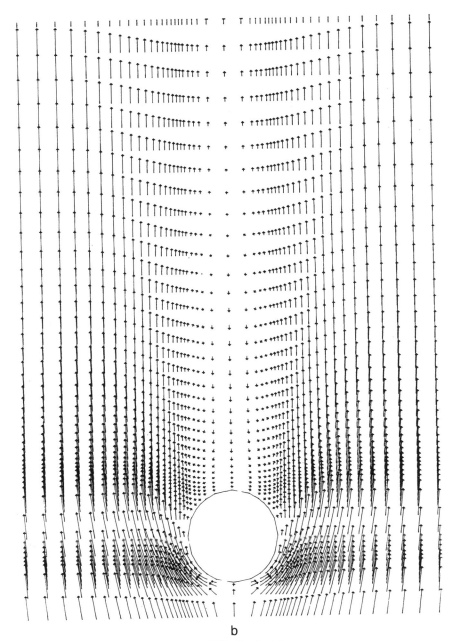

b

FIG. 3.19. *Continued.*

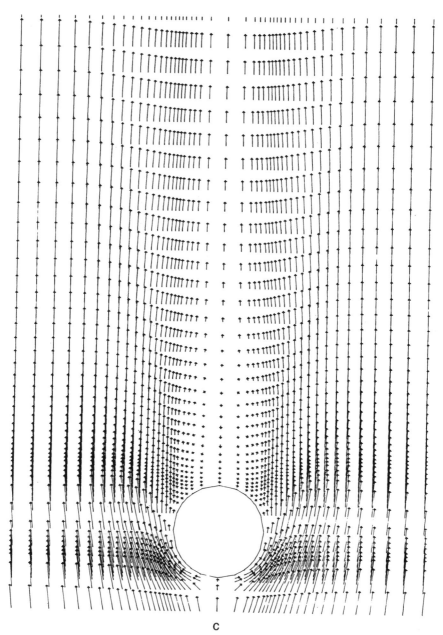

C

FIG. 3.19. *Continued.*

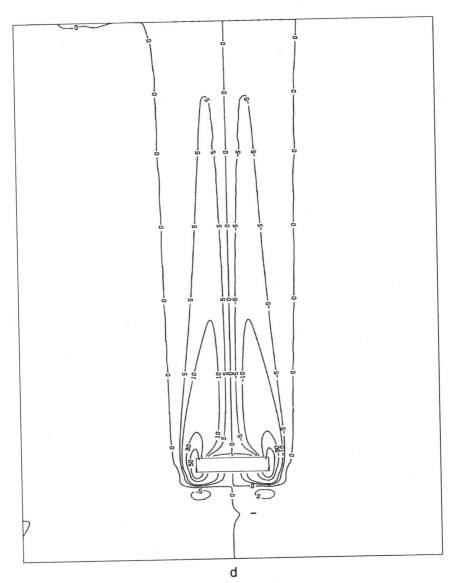

d

FIG. 3.19. *Continued.*

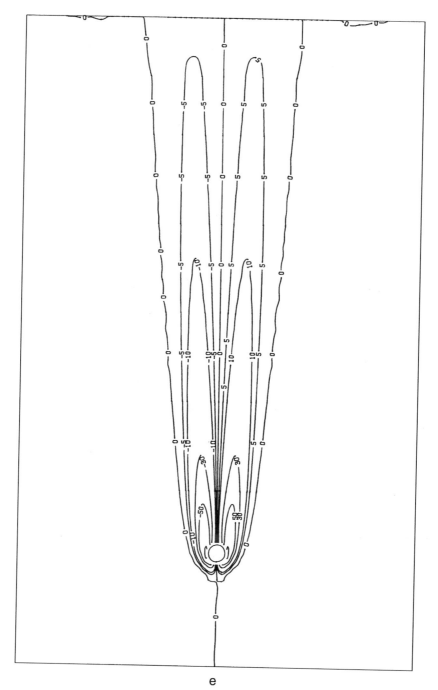

e

FIG. 3.19. *Continued.*

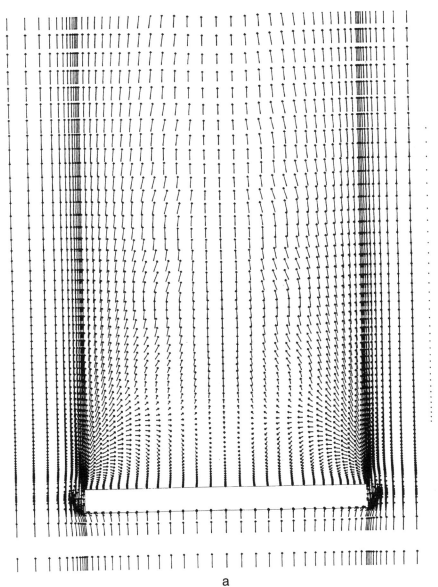

a

FIG. 3.20. Computed flow fields for flow past a short column at Re = 70. (a) Velocity field in the central cross section, broadside view. (b) Velocity field in the central cross section, end view. (c) Velocity field in a cross section near an end surface, end view. (d) Vorticity field in the central cross section, broadside view. (e) Vorticity field in the central cross section, end view.

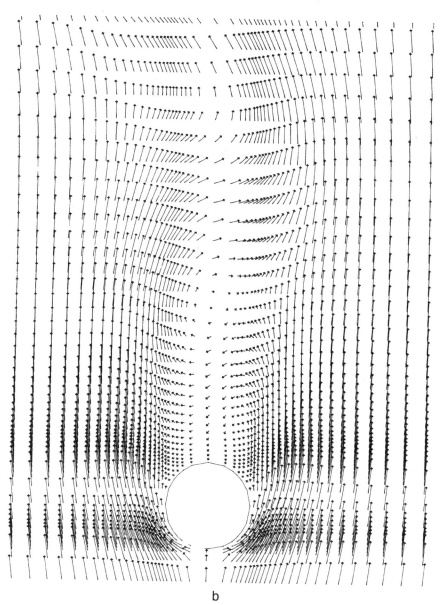

b

FIG. 3.20. *Continued.*

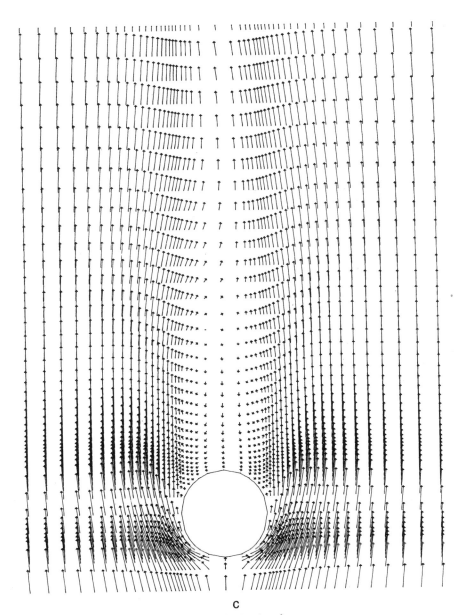

c

FIG. 3.20. *Continued.*

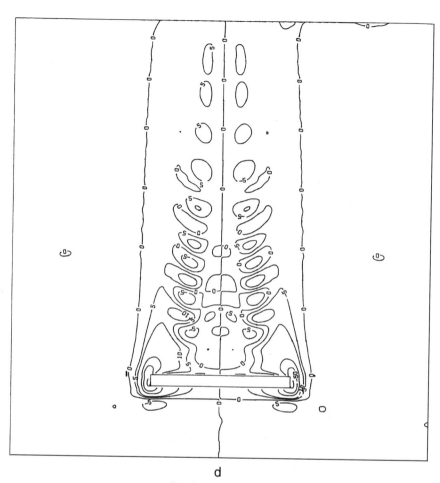

d

FIG. 3.20. *Continued.*

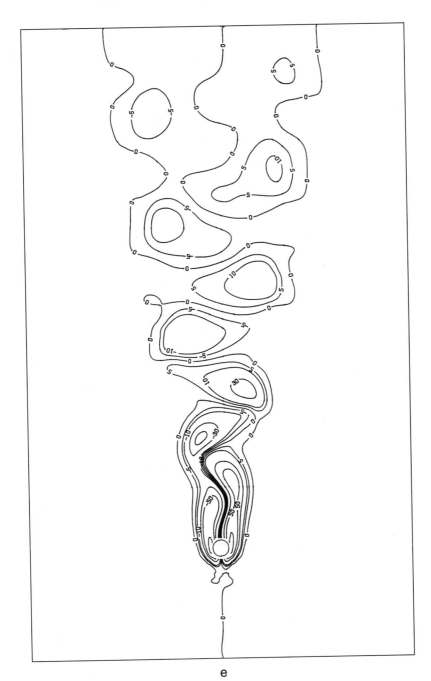

e

FIG. 3.20. *Continued.*

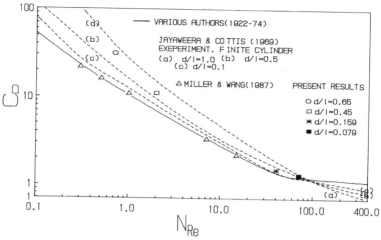

FIG. 3.21. Drag coefficients for flow past cylinders of various d/l ratio versus Re. The solid line and the triangles are for infinitely long cylinders. It is seen that as the Reynolds number increases, the cylinder becomes longer and the drag coefficient becomes closer to that of the infinitely long cylinder.

exhibits the same features as the tracer streaks in Figure 3.14b. Again, near the end the effect of the column on the flow becomes weak and the eddy feature becomes less pronounced.

Figures 3.20d and 3.20e show the vorticity fields at this instant. It is interesting to note that while the vorticity distribution in Figure 3.20d is symmetric with respect to the z-axis, that in Figure 3.20c is not. Obviously, the alternating nature of the shedding eddies is only revealed in the end view.

Figure 3.21 shows the comparison between the computed drag coefficients with those obtained by the other theoretical and experimental results. The drag coefficient is defined as

$$C_{\rm D} = \frac{D}{\rho V_\infty^2 \alpha}$$

(3.16)

where α is one-half of the cross-sectional area of the cylinder normal to the flow direction. Obviously, the drag coefficients of the present results differ from the results for infinite long cylinders. The difference is the greater the smaller the Reynolds numbers. This is due to the fact that the dimensions of the columns for lower Reynolds numbers are such that their shapes differ more from infinitely long cylinders. On the other hand, columns for higher Reynolds numbers are closer to the shape of infinitely long cylinders, hence their drag coefficients are closer to each other. This implies that the theoretical results obtained by previous investigators regarding the behavior of columnar ice crystals are probably reasonable for the case of larger ice columns, but not for short columns. The collection efficiencies of small droplets by and ventilation factors of falling ice columns computed based on a new scheme recently performed by us did show such trends and will be reported elsewhere. The discrepancies are mainly for Re < 10. It is seen here that the coefficients are practically the same as that of infinitely long cylinders for Re > 10. The drag coefficients calculated here can be fitted by the following expression:

$$\log_{10} C_D = 2.44389 - 4.21639A - 0.20098A^2 + 2.32216A^3 \qquad (3.17)$$

where

$$A = \frac{\log_{10} Re + 1.0}{3.60206} \qquad (3.18)$$

This formula is valid within the range $0.2 < Re < 100$. It fits the computed data to within a few percent. Note that in reality the drag coefficient is also a function of the aspect ratio of the cylinder, which is not explicitly represented in Eq. (3.17); hence, strictly speaking, this fit is only applicable to those cases indicated in Table 3.2. But judging from the smooth behavior of this relation, we feel that it is probably applicable to columnar crystals with dimensions satisfying Auer and Veal's (1970) relations and with flow Reynolds numbers in the aforementioned range. It would be desirable to find a relation of C_D as a function of the aspect ratio. However, more calculations are needed to establish this relation.

Figure 3.22 shows examples of the dimensionless pressure parameter K in the central section of the cylinder surface for Re = 40 and 70. K is defined as

$$K = (P - P_\infty)/(\rho V_\infty^2/2)$$

where P_∞ is the pressure far away from the cylinder. In the computation, it was obtained by taking the average of pressure values at the boundaries. There are no measured values available for 3-D flow past cylinders, so no experimental verification is possible at present. However, comparisons with pressures measured for 2-D flow cases may shed some light on the plausibility of our results. Thus when compared to the experimental results of Grove et al. (1964) for 2-D flow past a cylinder at Re = 40, the present results show similarities in shape, but slightly different in values. In particular, the minimum pressure (at $\theta \approx 70°$) in the present results is somewhat higher than the 2-D case. The result seems to be reasonable since a 3-D cylinder allows the fluid to pass the end surfaces to reach the wake, hence higher rear pressures. For a cylinder of the same dimension (i.e., the short cylinder) but with Re = 70, the pressures in the rear are higher, in agreement with the general behavior of 2-D flows. (Note that the rear pressures for Re = 70 here represent time-averaged values.) On the other hand, the rear pressures of a longer cylinder (Re = 70) are lower than that of a shorter cylinder. This also seems to be reasonable since a longer cylinder is closer to a 2-D cylinder, hence lower rear pressures.

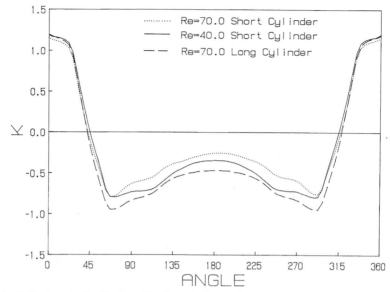

FIG. 3.22. Angular distribution of the dimensionless surface pressure parameter K in the central yz (end view) cross section.

Since the cylinders are finite, surface pressures at other cross sections are different from that in the central cross section. Figures 3.23a and 3.23b show the surface pressure contours for Re = 40 and 70, respectively. As expected, the effect of the end surface is most pronounced near the end, but unimportant near the center portion of the cylinder. Again, the distribution over the longer cylinder (Re = 70) is closer to that over an infinitely long cylinder. In the latter case, the contours would be simply parallel straight lines.

Figure 3.24 shows the surface vorticity distributions for the Re = 40 and 70 cases. Like the pressure distributions, the angular distribution of surface vorticity in the central cross-section, is close to the 2-D flow case. The peak absolute values, however, are slightly lower than the 2-D cases for both Reynolds numbers as reported by Braza *et al.* (1986). Figures 3.25a and 3.25b show the surface vorticity contours. Near the end surfaces, the vorticity distributions are somewhat different due to the strong end effect. Otherwise the vorticity distributions are similar to the 2-D cases.

Finally, the pressure distribution of the whole domain for the case of Re = 70 is shown in Figures 3.26a–c as an example of the presence of the ice column on the overall pressure field. The common features of front stagnation high pressure and the low pressures in the wake region are well illustrated here. The cases of steady flows would have similar, but simpler, features. The pressure distribution around the falling crystal has an impact on the collision efficiencies of other particles with ice crystals, as has been explained by Pitter *et al.* (1974).

3.5.3. General Features of the Flow Fields around Falling Hexagonal Ice Plates

Pioneering numerical work on the flow fields around falling ice plates was performed by Pitter *et al.* (1973), who used thin oblate spheroids to approximate planar ice crystals. In the present study, we use the actual hexagonal shape to model the ice plates whose dimensions are given in Table 3.3. The corresponding range of the Reynolds numbers is from 1 to 20. However, additional cases were also computed as needed to demonstrate the flow fields.

Examples of steady flow fields around hexagonal crystals are shown in Figures 3.27 and 3.28 which represent cases of Reynolds numbers 2 and 20, respectively. These are steady flow cases. The flow fields look similar to those obtained by Pitter *et al.* (1973).

The flow field of Re = 1 (figure not shown) does not indicate the existence of standing eddies. But there are already standing eddies formed in the wake region of the crystal at Re = 2. This is consistent with Pitter *et al.* (1973) who indicated that the eddies start to appear at Re = 1.5. As expected, the eddies become larger at higher Reynolds numbers.

Experiments of Willmarth *et al.* (1964) showed that at Re ≥ 100, eddy shedding occurs in the downstream of a falling disk. Such unsteady behavior can be simulated

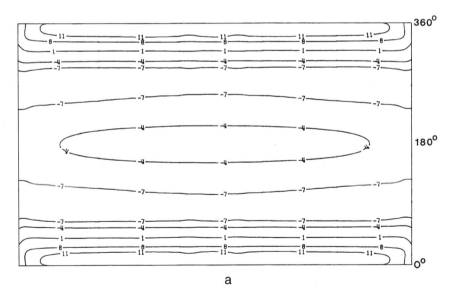

a

FIG. 3.23. Surface pressure distribution for flow past a finite cylinder. (a) Re = 40; (b) Re = 70.

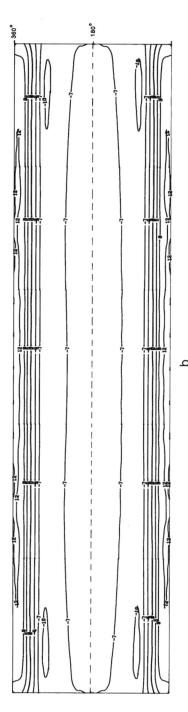

FIG. 3.23. Continued.

Surface Vorticity on the Central Cross-section

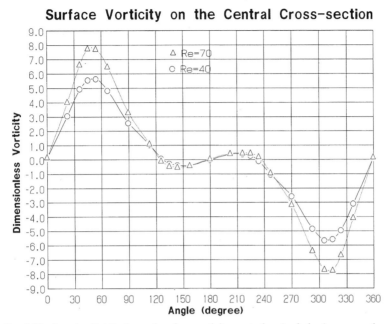

FIG. 3.24. Angular distributions of surface vorticity, central yz (end view) cross section.

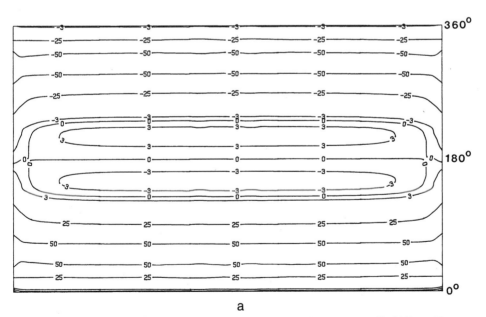

a

FIG. 3.25. Surface vorticity contours for flow past finite cylinders. (a) Re = 40; (b) Re = 70.

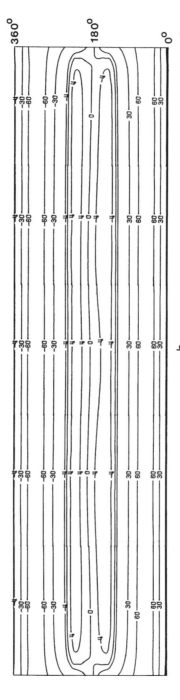

FIG. 3.25. *Continued.*

using the same technique as we did for the finite cylinders. Figure 3.29 shows the simulated unsteady flow fields for flow past hexagonal plates at Re = 140 at four different time steps after the steady-state solution is achieved. The initial perturbation introduced (after the steady-state solution has been obtained) was again $0.3V_{\infty}$, which has been proven to be adequate for kicking up the shedding. It can be seen that the flow field is obviously asymmetric due to the shedding.

a

FIG. 3.26. Computed pressure fields for flow past a finite cylinder at Re = 70. (a) Broadside view, central cross section. (b) Pressure field in the central cross section, end view. (c) Pressure field in a cross section near an end surface, end view.

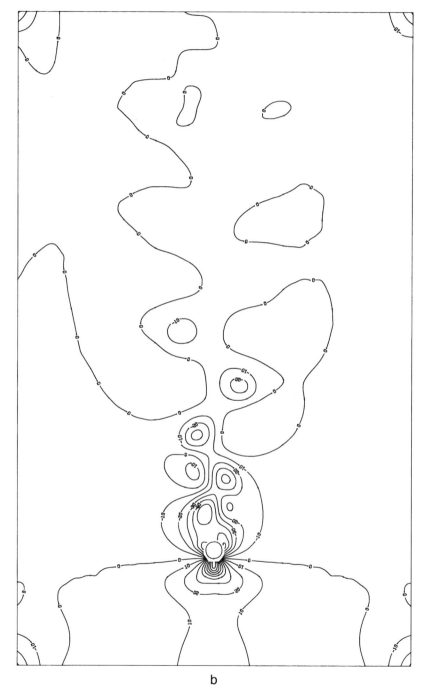

b

FIG. 3.26. *Continued.*

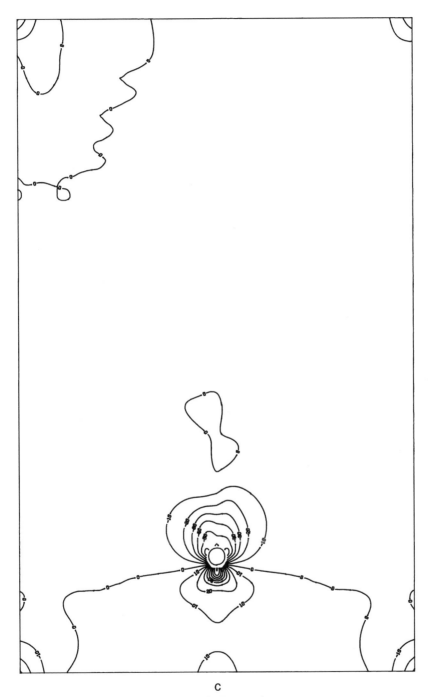

C

FIG. 3.26. *Continued.*

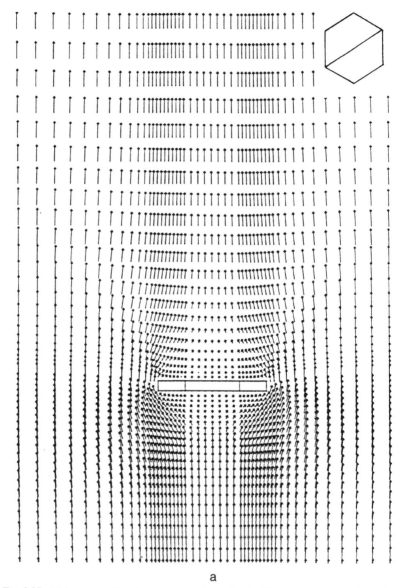

a

FIG. 3.27. (a) A cross-sectional view of the velocity fields of flow past a hexagonal ice plate at Re = 2. The cross section is indicated by the line marked in the crystal in the upper right corner of the figures. Vectors represent projections of the 3-D vectors onto that cross-sectional plane. (b) Same as Fig. 3.27a but for another cross section.

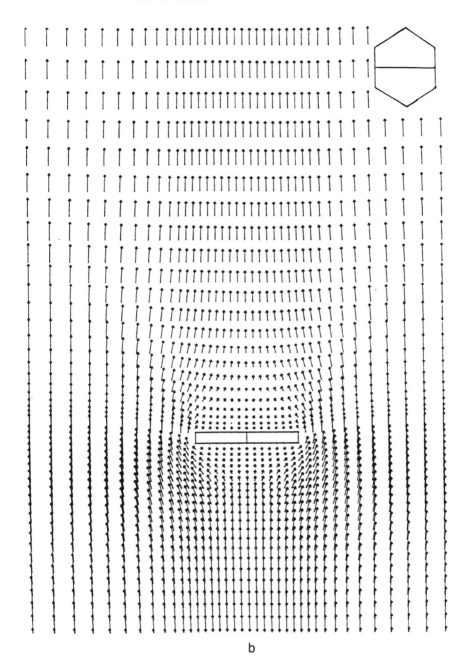

b

FIG. 3.27. *Continued.*

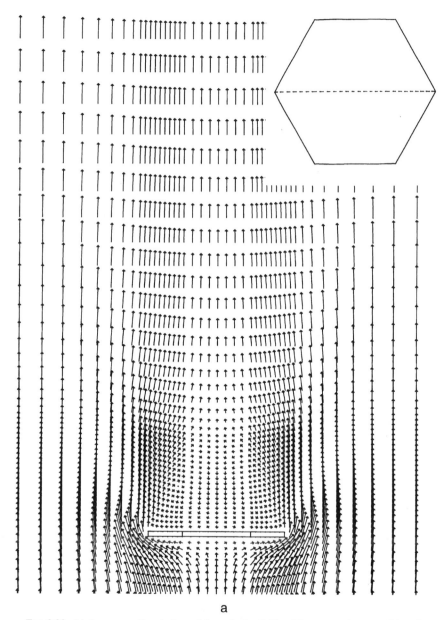

a

FIG. 3.28. (a) A cross-sectional view of the velocity fields of flow past a hexagonal ice plate at Re = 20. The cross-section is indicated by the dashed line in the upper right corner of the figures. Vectors represent projections of the 3-D vectors onto that cross sectional plane. (b) Same as Fig. 3.28a but for another cross section. (c) Same as Fig. 3.28a but for another cross section.

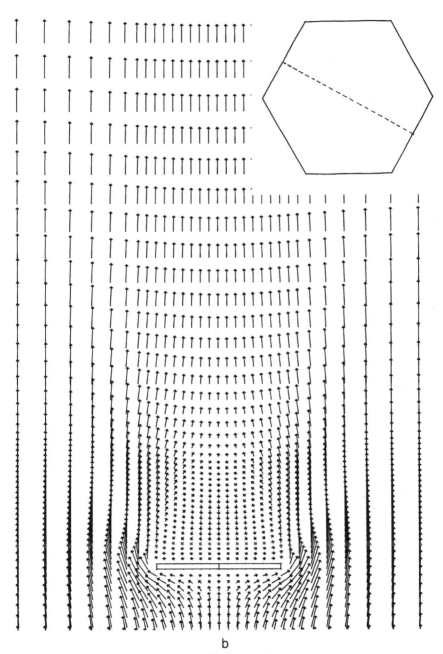

b

FIG. 3.28. *Continued.*

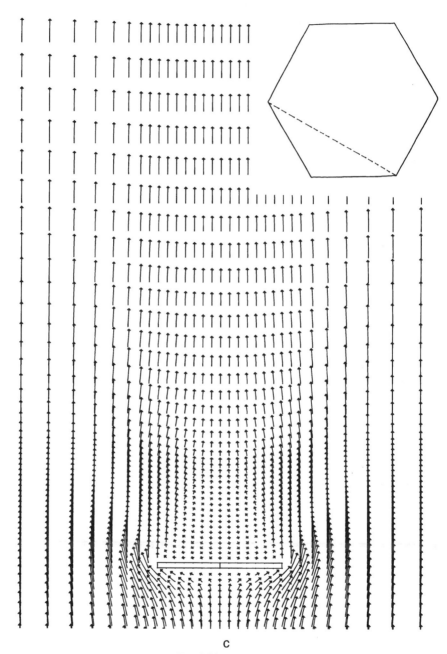

C

FIG. 3.28. *Continued.*

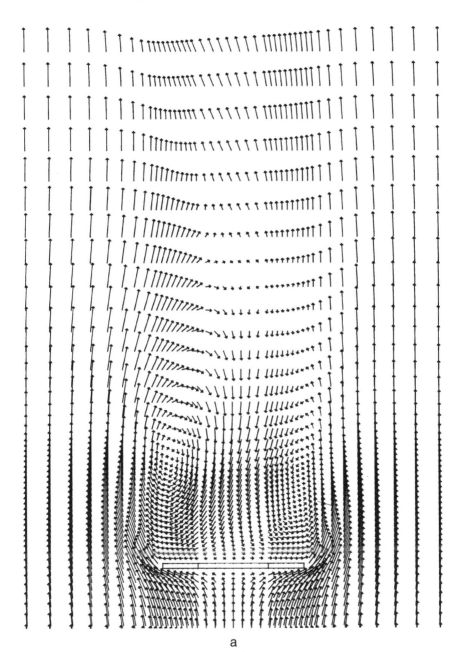

a

FIG. 3.29. The unsteady flow field past a hexagonal ice plate at Re = 140. (a) $t = 159$;
(b) $t = 162$; (c) $t = 168$; (d) $t = 180$.

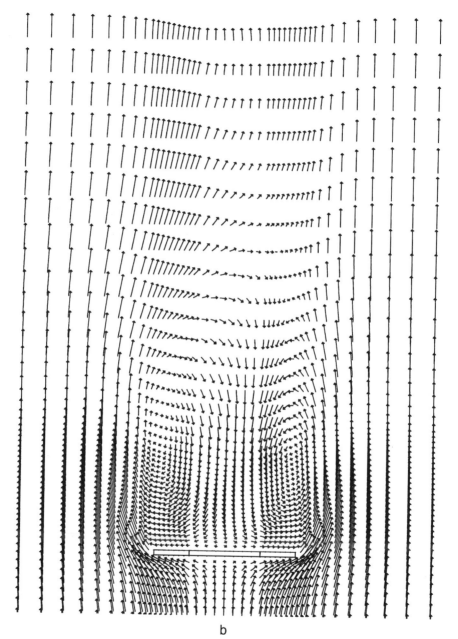

b

FIG. 3.29. *Continued.*

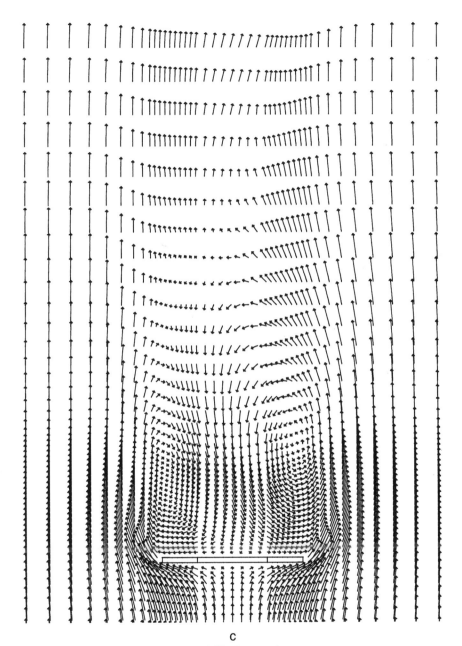

c

FIG. 3.29. *Continued.*

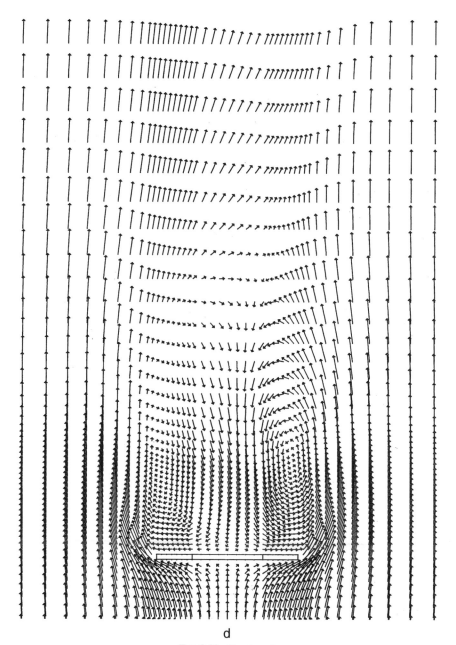

d

FIG. 3.29. *Continued.*

Detailed analysis of how shedding starts has not been done yet, but it is expected that the shedding would start at a particular corner and the point of detachment would rotate around on the plate.

It must be stressed here that the flow fields described above are computed assuming that the plate position is fixed with respect to the incoming air flow, that is, the angle between the c-axis (normal to the plate basal surface) of the plate and the general flow is kept at $90°$. In reality, falling plates are known to perform zigzag motion, which implies that the angle is not constant, but is actually a function of time. In order to simulate such zigzag motions, one must use very small time steps for adequate accuracy. Owing to the constraint of computing resources, these cases are not yet simulated here, but we are currently studying them. However, it is expected that the above results should give good approximations, especially when the variations of the angles are not large.

No experimental measurements appear to be available for flow properties past hexagonal plates. The results of Willmarth et al. (1964) for flow past circular plates are the closest cases for the comparison purpose. But here the comparison is difficult to make because the aspect ratios of the computed and experimental results are different. For that same reason the comparison between our results and those of Pitter et al. (1973) is also difficult to make. The aspect ratio of the plates calculated here varies with the Reynolds number whereas the thin oblate spheroids in Pitter et al. (1973) have fixed aspect ratios ($h/d = 0.05$), therefore, the computed results cannot be compared directly except for the case of Re = 20. At this Reynolds number, the drag coefficient obtained by Pitter et al. (1973) agrees with the present result for hexagonal plate to within 1%. The results of Willmarth et al. (1964) do not have the case of Re = 20, and the two cases that are close have rather different aspect ratios ($h/d = 0.0033$ for Re = 15.7 and 0.00167 for Re = 29.1). In other Reynolds number cases, the differences are somewhat larger, possibly due to the different aspect ratios (and, of course, somewhat different shapes). But even there, the largest error which occurs at Re = 140 is less than 15%. Thus it seems to be fair to conclude that the general trend and the magnitude of C_D constants are indeed quite similar for the present and Pitter et al. (1973) cases, and hence the predictions made by Pitter et al. (1973) seem to be generally valid.

Figure 3.30 shows the massless tracer streaks computed for flow past a hexagonal plate at Re = 240. Although no exact comparison can be made with experimental photographs, the main features of the streaks are very similar to the pictures taken by Willmarth et al. (1964, their Fig. 7) for a falling Plexiglas disk at Re $\approx$ 100 (but in a disturbed medium).

3.5.4. General Features of the Flow Fields around Falling Broad-Branch Crystals

Broad-branch crystals are also a common form of ice and snow crystals. We have not seen any quantitative measurements or calculations about the flow fields

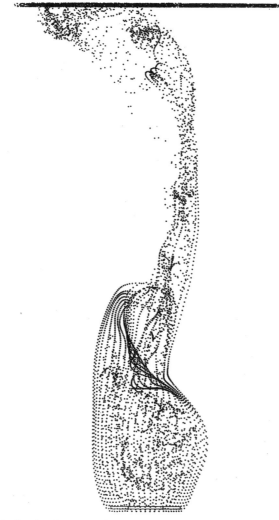

FIG. 3.30. Streaks of tracer particles for flow past a hexagonal plate at Re = 240.

around a falling broad-branch crystal, so the present results may be the first of their kind. Needless to say, it would be desirable to have experimental measurements in the future to compare with our computational results.

Because broad-branch crystals are basically planar crystals, as are the ice plates discussed in the previous section, the flow fields around them are expected to be similar (albeit different in magnitudes) to that around ice plates. This is indeed the case, as shown by the present calculations. The major difference between a plate and a broad-branch crystal, of course, is in the gaps between the branches of the latter. Figures 3.31 and 3.32 show the computed flow fields around falling broad-branch crystals for Re = 2.0 and 20, respectively. There are already standing eddies in the flow field of Re = 2.0 although they are very small, just barely discernible. The size of the eddies is smaller than that for flow past plates at the same Reynolds number. This may be understood by noting that the gaps between branches would allow the fluid to go through more easily and therefore reduce the tendency of creating return flow which constitutes the eddies.

The flow in the gap region is of particular interest since this phenomenon has never been studied before. Figure 3.33 shows a special cross section of the flow field that reveals the nature of this regional flow. The flow converges slightly before entering the gap region, becomes relatively straight in the gap, and then diverges slightly upon leaving it. The magnitude of the flow velocity is relatively small.

The standing eddy size again increases with increasing Reynolds number as shown by Figure 3.34 for Re = 20. The same convergent–straight–divergent behavior occurs in the gap region, as shown in Figure 3.33, but it is more evident because of the slightly higher velocity than the Re = 2.0 case. However, the velocity in this region is still small when compared to the general flow. This generally small velocity phenomenon seems to indicate that the flow in the gap regions can be approximated as creeping flow. If this observation can be sustained by future studies, then it implies that we can use the creeping flow theory to treat the even more intricate case of flow going through the dendritic crystals where the gaps are even smaller.

Finally, Figure 3.35 shows the variation of drag coefficients as a function of Reynolds number for flow around hexagonal plates and broad-branch crystals. The drag coefficient for a broad-branch crystal is greater than that for a plate at the same Reynolds number. The drag coefficients for very thin oblate spheroids as calculated by Pitter *et al.* (1973) are also plotted for comparison. The drag coefficients for the hexagonal plates and broad-branch crystals computed in the present study can be fitted by the following empirical formulas which use a functional form similar to that of Pitter *et al.* (1973) for circular plates:

$$C_D = \left(\frac{64}{\pi \mathrm{Re}} \right)(1 + 0.078\ \mathrm{Re}^{0.945}) \tag{3.19}$$

$$C_D = \left(\frac{64}{\pi \mathrm{Re}} \right)(1 + 0.142\ \mathrm{Re}^{0.887}) \tag{3.20}$$

These formulas are valid for the range of Re between 0.2 and 150. Both fit the calculated values of drag coefficients to within 1.5%. Currently, no experimental data are available to verify the calculations of flow fields for flow past broad-branch crystals.

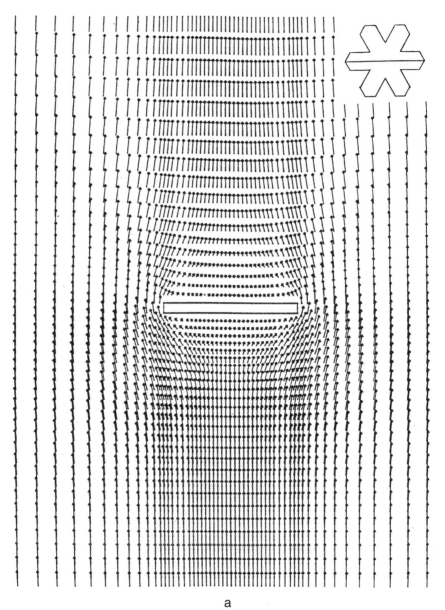

a

FIG. 3.31. (a) Velocity fields of flow past a broad-branch ice crystal at Re = 2. The cross section is indicated by the line marked in the crystal in the upper right corner of the figures. (b) Same as Fig. 3.31a but for another cross section.

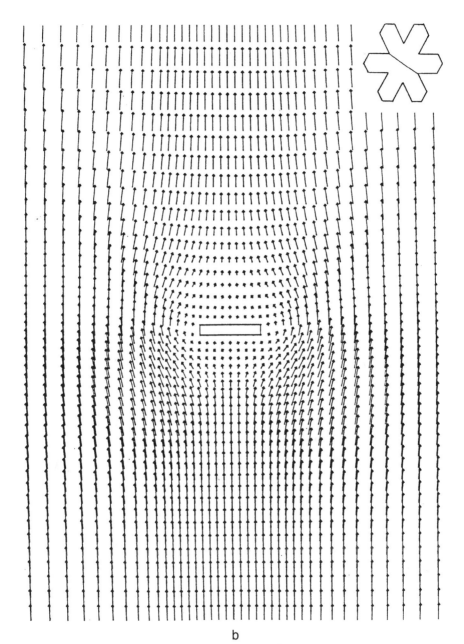

b

FIG. 3.31. *Continued.*

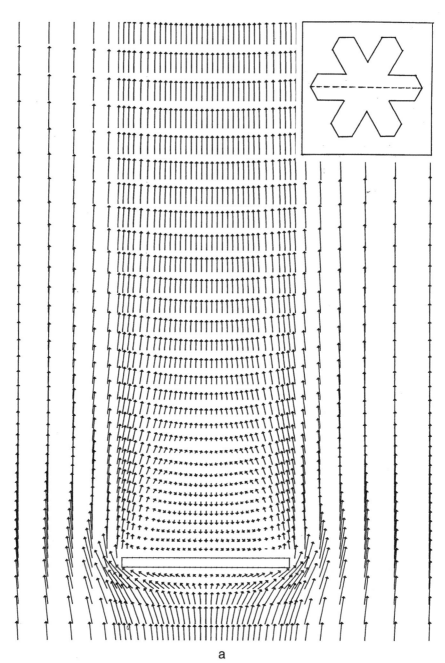

a

FIG. 3.32. (a) Velocity fields of flow past a broad-branch ice crystal at Re = 20. The dashed line in the upper right corner of the figure indicates the cross section. (b) Same as Fig. 3.32a but for another cross section.

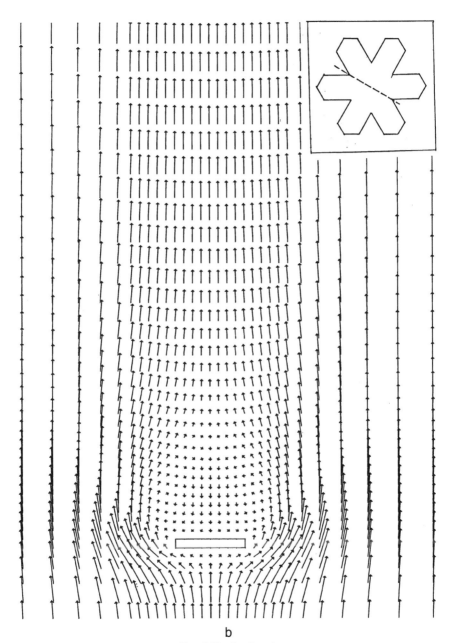

b

FIG. 3.32. *Continued.*

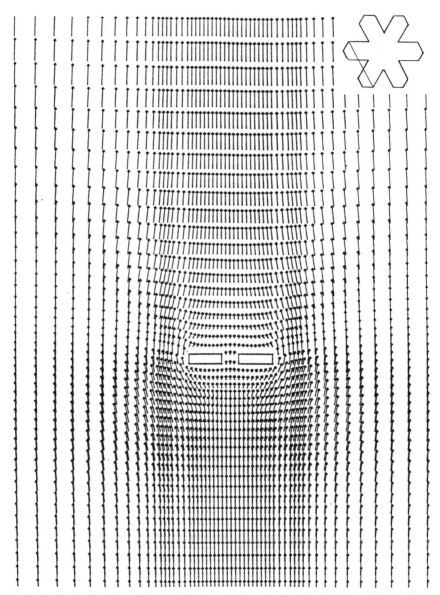

FIG. 3.33. Velocity field of flow past a branch gap of a broad-branch ice crystal at Re = 2. The line in the upper right corner of the figure indicates the cross section.

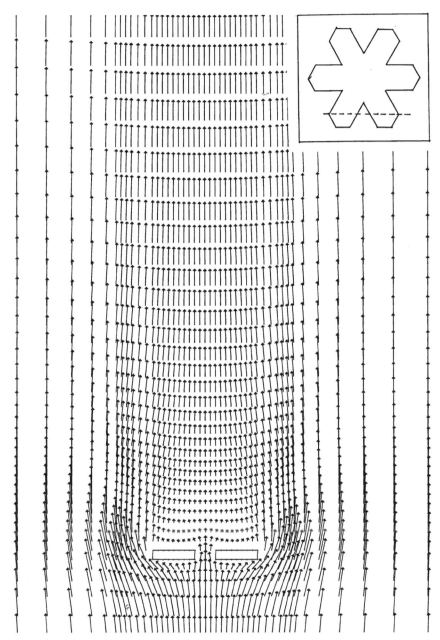

FIG. 3.34. Velocity field of flow past a branch gap of a broad-branch ice crystal at Re = 20. The dashed line in the upper right corner of the figure indicates the cross section.

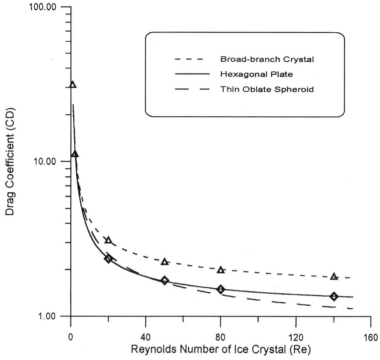

FIG. 3.35. Drag coefficients as a function of Reynolds number for flow past hexagonal plates and broad-branch crystals.

3.5.5. Remarks on the Flow Fields around Larger Falling Ice Particles

In this study we showed that the unsteady flow fields around three types of falling ice crystals can be determined by numerically solving the time-dependent Navier–Stokes equations. Nonuniform Cartesian grids were used owing to the complexity of the crystal geometry which always results in three-dimensional flow fields. Comparisons of current numerical results with experimental data, when available, show good quantitative agreement between the two. This indicates that the present technique is capable of realistically simulating the flow fields around falling ice crystals. This also gives us some confidence in using the computed flow fields to derive other cloud physical quantities that depend strongly on hydrodynamic effects such as the collision efficiencies of small particles (cloud droplets, aerosol particles, etc.) with these ice crystals and the ventilation factors that influence the evaporation, cooling, and diffusional growth of these crystals. The computations of these quantities were also performed by us and will be reported in the next chapter.

It is necessary to remind the reader that the present results represent only the first attempt to simulate the flow fields around falling ice crystals and hence are far from perfect. In particular, the ice crystals in the present study are all held fixed with respect to air flow; thus the zigzag motion of some larger crystals is not simulated. In addition, no theoretical computations on the flow fields around large hydrometeors such as graupel, hail, and raindrops have been reported, although some preliminary tests on the computation of flow past conical hydrometeors have been recently attempted by the author's group. These tests were done using the conical shapes described by the expression in Eq. (2.25), and preliminary results show that they can be done in the similar manner as reported here. But, of course, more work is needed to obtain final results. We are hoping that this deficiency can be removed in the near future. Considerable computing resources will be necessary for this type of study since the required time step to achieve reasonable accuracy is likely to be rather small. However, judging from the recent impressive advances of computer speed, such computations will probably become possible soon.

Finally, we lack experimental measurements of flow fields around falling ice crystals, which are necessary to verify the theoretical results. Ideal equipment for this purpose would be a vertical wind tunnel such as the one used by Beard and Pruppacher (1971) or Mitra *et al.* (1990). Presumably, measurements of this kind will become available in the future.

4. VAPOR DIFFUSION, VENTILATION, AND COLLISIONAL EFFICIENCIES OF ICE CRYSTALS

4.1. Introduction

The development of ice-containing clouds is naturally influenced greatly by the growth of individual ice particles. There are two main modes of ice particle growth in clouds: diffusion and collision. Diffusional growth is the mechanism by which water vapor molecules are added directly to the ice particles. The opposite of this is, of course, evaporation or sublimation whereby water molecules leave the surface of ice particles and become vapor. I shall use the generic term "evaporation" to represent the vaporization of both liquid and solid phases. Collisional growth is the mechanism by which ice particles become bigger by collecting either water droplets or other ice particles. We are only concerned with the collision between ice particles and supercooled droplets, a process called *riming*. The collection and melting of ice particles by water drops or the collection of ice particles by ice particles (aggregation process) will not be considered here.

Both diffusional growth and collisional growth of falling ice particles are certainly influenced by the flow fields around the ice particles. The flow fields we obtained in the previous section play a vital role in determining the rates of these two growth modes. In the case of diffusional growth, the flow field influences the vapor distribution around the ice particle that results in a general enhancement of the vapor flux toward (or away from, in the case of evaporation) the ice crystal surface, a phenomenon called the *ventilation effect*. This effect makes a falling crystal grow (or evaporate) faster than when it is stationary, relative to the air. In the following, we first treat the vapor diffusion problem around an ice column to demonstrate some analytical solutions, then we examine the ventilation problem.

The other main topic is the collisional efficiency of ice crystals with supercooled droplets. The presence of the flow field due to the motion of the ice crystal in air affects the droplet and makes the collision efficiency differ significantly from 1 (i.e., the geometric collision efficiency). To determine this collision efficiency, we performed calculations to determine the trajectories of the droplet under the influence of the hydrodynamic drag due to the flow field. The details are reported in Section 4.4. The treatments below follow mainly Ji and Wang (1998) and Wang and Ji (1997).

4.2. Vapor Diffusion Fields around a Stationary Columnar Ice Crystal

The rate of diffusional ice crystal growth depends on the magnitudes of water vapor density gradient at the surface of the crystal. Thus, the ideal way to solve the theoretical problem of diffusional growth of ice crystals would be to obtain the vapor density distribution around the crystal first, and then determine the vapor density gradients and integrate the vapor flux density (diffusivity times gradient) over the crystal surface to obtain the growth rate. However, in order to determine the vapor density distribution, one would need to specify the boundary conditions. This can be done easily for the outer boundary (the vapor density is a constant far away from the crystal surface), but is not so easy for the inner boundary (the vapor density is another constant on the crystal surface). The main problem is the difficulty in describing the crystal surface by simple mathematical expressions (and this has only recently been partially achieved by Wang, 1999, as described in Section 2) because most ice crystals consist of mixed surfaces, hence it is necessary to prescribe the boundary conditions in a "mixed" way. The boundaries between these mixed surfaces are usually discontinuous, and mathematical treatments here become very complicated.

For this reason, the conventional way of treating the diffusional growth of ice crystals is to sidestep the solution of vapor density distribution and go on to determine the growth rate by means of the so-called electrostatic analog, which utilizes the Gauss law in electrostatics and the relation between electric potential (analogous to vapor density), electric charge (analogous to the growth rate), and capacitance. For details of this method, see Pruppacher and Klett (1997).

However, using the electrostatic analog requires that we know the capacitance of the ice crystal, and the capacitance is usually unknown unless measured experimentally. Furthermore, there are occasions when knowledge of the vapor density distribution is desirable (such as the direction of the crystal growth), but that cannot be provided by the electrostatic analog.

In the following section, the formal way of determining the vapor distribution around a columnar ice crystal (approximated as a finite cylinder) is demonstrated as an example for treating general vapor diffusion problems around ice crystals. This example consists of mixed boundaries. It is hoped that this difficulty can be removed in the future by using the expression developed in Section 2 for describing crystal surfaces.

4.2.1. Mathematical Formulation

The word "potential" used in the following can represent many physical quantities, including vapor density, temperature, electric potential, and so on. The reason is that the distributions of these quantities in a "source-free" space can be described by the Laplace equation, and hence the solutions are the same if they have the same boundary condition (which they do).

In this section, we investigate only columnar crystals. The true shape of these columns is hexagonal, which is very difficult to treat. We shall approximate these columns by circular cylinders of equal diameter and length. This approximation should represent an improvement to the prolate spheroid approximation, which lacks sharp edges at both ends. Even with such a simplification, the present problem is still not simple. The main difficulty here is that such a cylinder consists of two types of boundary surfaces: the cylindrical side surface and the planar end surfaces. This poses a mixed boundary problem, and the usual orthogonal function techniques become hopelessly complicated. In the present study, we applied a technique developed by Smythe (1956, 1962), who used the series expansion and integral transformation methods to determine the charge density distribution and capacitance of a charged right circular cylinder. Here we extend his method to derive an analytical expression for the potential distribution, which can represent any one of the electric potential, temperature, or vapor density fields surrounding such a cylinder. In the following subsection we formulate and calculate the electrostatic fields. Conversions of temperature and vapor density fields are given at the end.

Figure 4.1 shows the configuration of the cylinder and the nomenclature of the problem. Our task is to determine the field strength at any outside point P (a_1, ϕ_1, z_1). We shall assume that the cylinder is conducting and that it is charged to surface potential V_0. This may be valid for a single crystalline ice column. The charge densities on the side and end surfaces can be expressed as (Smythe, 1956; Wang et al., 1985)

$$\sigma_s = \sum_{n=0}^{\infty} A_n \left(b^2 - z_0^2\right)^{n-1/3} \tag{4.1}$$

and

$$\sigma_e = \sum_{n=0}^{\infty} B_n \left(1 - \rho_0^2\right)^{n-1/3} \tag{4.2}$$

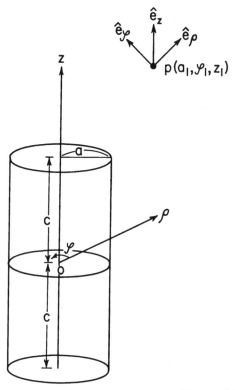

FIG. 4.1. Nomenclature of the problem. The radius and half-length of the cylinder are a and c, respectively; ρ is the radial and z is the vertical coordinate.

respectively, where b, z_0, and ρ_0 are dimensionless quantities defined by

$$b = c/a, \qquad z_0 = z/c, \qquad \rho_0 = \rho/a \qquad (4.3)$$

where a is the radius, c is the half-length of the cylinder, and ρ and z are the respective radial and vertical coordinates. A_n and B_n are coefficients to be determined. The above charge distributions are such that the following equations are satisfied on a closed surface enclosing the origin:

$$\left[\frac{d^{2p} V(z, 0)}{dz^{2p}}\right]_{z=0} = \begin{cases} V_0 & \text{when } p = 0, \\ 0 & \text{when } p \neq 0 \end{cases} \qquad (4.4)$$

where V is the potential.

Very near the edge, as indicated by Smythe (1956), the charge distributions (4.1) and (4.2) become equal and both approach the charge density on a rectangular wedge which is proportional to $\delta^{-1/3}$ where δ is the distance from the edge. The property of such a wedge is given in Smythe (1962).

The next step is to determine the coefficients A_n and B_n. This is done first by calculating the potential on the axis of the cylinder due to the charge densities of Eqs. (4.1) and (4.2). The expression of the potential is then differentiated with respect to z_0, and finally z_0 is equated to 0 (see Smythe, 1962, for details). The result is

$$\frac{d^{2p}V}{dz_0^{2p}} = \sum_{n=0}^{\infty} \left[\left(\frac{MB_n}{n + \frac{2}{3}} \right) F_1 + MNGA_n F_2 \right] \tag{4.5}$$

where

$$M = \frac{\frac{1}{2}a(2p)!}{\varepsilon(1+b^2)^{p+1/2}}, \qquad N = \frac{b^{2n+1/3}\left(n - \frac{1}{3}\right)!\left(p - \frac{1}{2}\right)!}{(-1)^p 2^{2/3}\left(n + \frac{1}{6}\right)!p!},$$

$$G = 0.674463408 \tag{4.6}$$

and

$$F_1 = F_1\left[p + \tfrac{1}{2},\, n - p + \tfrac{2}{3};\, n + 1 + \tfrac{2}{3};\, (1+b^2)^{-1}\right]$$
$$F_2 = F_2\left[p + \tfrac{1}{2},\, n - p + \tfrac{2}{3};\, n + 1 + \tfrac{1}{6};\, b^2(1+b^2)^{-1}\right] \tag{4.7}$$

are two hypergeometric functions. Equation (4.5) can be substituted into Eq. (4.4). This yields a number of simultaneous equations which can be used for solving coefficients A_n and B_n. But we still need an extra condition to relate A_0 to B_0. This additional condition is

$$A_0 = b^{1/3}B_0 \tag{4.8}$$

so that the charge distributions σ_s and σ_e match at the edge of the cylinder. Thus by solving a number of simultaneous equations of the type of Eq. (4.5) plus Eq. (4.8), the coefficients A_n and B_n can be determined. The number of A_n and B_n required to achieve good accuracy (so that the correct surface potential is reproduced) need not be large, usually being six to eight terms.

We now derive an expression for the electric potential outside the cylinder based on the charge density distributions (4.1) and (4.2). We first have to determine the distance r from a point $B(a, \phi, z)$ on the cylinder to the field point $P(a_1, \phi_1, z_1)$. This is (see Fig. 4.1)

$$r = \left[(z - z_1)^2 + a^2 + a_1^2 - 2aa_1\cos(\phi_1 - \phi)\right]^{1/2} \tag{4.9}$$

The contribution of the charges on the cylindrical side surface to the potential at P is therefore

$$
\begin{aligned}
V_s(a_1, \phi_1, z_1) &= \int_s \frac{\sigma_s}{4\pi\varepsilon r}\, ds \\
&= \int_s \frac{\sum A_n[(c^2 - z^2)/a^2]^{n-1/3} a\, d\phi\, dz}{4\pi\varepsilon[(z_1 - z)^2 + a^2 + a_1^2 - 2aa_1\cos(\phi_1 - \phi)]^{1/2}}
\end{aligned}
\tag{4.10}
$$

where $ds = a\, d\phi\, dz$ is a surface element on the side surface. The angular part can be readily integrated to give

$$
V_s(a_1, \phi_1, z_1)
$$

$$
= \frac{1}{4\pi\varepsilon}\left\{ 2a \int_{-c}^{c} \sum A_n \left(\frac{c^2 - z^2}{a^2}\right)^{n-1/3} \frac{2}{\sqrt{\alpha_1^2 + \beta_1^2}} K\left(\sqrt{\frac{2\beta_1^2}{\alpha_1^2 + \beta_1^2}}\right) dz \right\}
\tag{4.11}
$$

where the complete elliptic integral K can be expressed as

$$
K(k) = \frac{\pi}{2}\left[1 + \left(\frac{1}{2}\right)^2 k^2 + \left(\frac{1\cdot 3}{2\cdot 4}\right)^2 k^4 + \left(\frac{1\cdot 3\cdot 5}{2\cdot 4\cdot 6}\right)^2 k^6 + \cdots\right], \qquad k^2 \le 1
\tag{4.12}
$$

and

$$
\begin{aligned}
\alpha_1^2 &= (z_1 - z)^2 + a^2 + a_1^2 \\
\beta_1^2 &= 2aa_1
\end{aligned}
\tag{4.13}
$$

By a similar consideration, the contributions of the charges on the upper and lower planar end surfaces to the potential at P are, respectively,

$$
V_{e,u}(a_1, \phi_1, z_1) = \int_s \frac{\sigma_e}{4\pi\varepsilon r}\, ds = \frac{1}{4\pi\varepsilon}\left\{ 2\int_0^a \left[\sum B_n \left(\frac{a^2 - \rho^2}{a^2}\right)^{n-1/3}\right] \right.
$$

$$
\left. \times \frac{2\rho}{\sqrt{\alpha_2^2 + \beta_2^2}} K\left(\sqrt{\frac{2\beta_2^2}{\alpha_2^2 + \beta_2^2}}\right) d\rho \right\}
\tag{4.14}
$$

and

$$
V_{e,l}(a_1, \phi_1, z_1)
$$

$$
= \frac{1}{4\pi\varepsilon}\left\{ 2\int_0^a \left[\sum B_n \left(\frac{a^2 - \rho^2}{a^2}\right)^{n-1/3}\right] \frac{2\rho}{\sqrt{\alpha_3^2 + \beta_2^2}} K\left(\sqrt{\frac{2\beta_2^2}{\alpha_3^2 + \beta_2^2}}\right) d\rho \right\}
\tag{4.15}
$$

where

$$\alpha_2^2 = (z_1 - c)^2 + a_1^2 + \rho^2$$

$$\beta_2^2 = 2a_1\rho \tag{4.16}$$

$$\alpha_3^2 = (z_1 + c)^2 + a_1^2 + \rho^2$$

The electric potential at an external point $P(a_1, \phi_1, z_1)$ due to the charged circular cylinder is therefore the sum of the above three parts (4.11), (4.14), and (4.15), i.e.,

$$V_P = V_s + V_{e,u} + V_{e,l} \tag{4.17}$$

The integrals (4.11), (4.14), and (4.15) are to be calculated by numerical methods.

Once the potential profiles are known, the electric fields can be calculated by taking the gradients of the potentials. Here, these gradients are taken numerically. One can, of course, determine the fields graphically. It is most convenient to use spline interpolation to find the neighboring V values once certain base point potentials are known. It is found that the potentials determined this way are very close to the values calculated using the full expressions. The difference is on the order of only 1%.

4.2.2. Examples

Potential fields were calculated for eight cylinders, corresponding to the eight columnar ice crystals in Schlamp et al. (1975) and Wang and Pruppacher (1980a). The ratios of half-length to radius (c/a) used in the calculations are 1.21, 1.43, 1.54, 1.67, 2.22, 3.33, 5.00, and 8.33, with approximate Reynolds numbers 0.2, 0.5, 0.7, 1.0, 2.0, 5.0, 10.0, and 20.0, respectively. Some examples are presented here. All results shown are in SI units. Figures 4.2 and 4.3 show the computed charge density distributions (in C/m^2) on the planar end surfaces and cylindrical side surfaces, respectively. All the cylinders are charged to a surface potential of 1 V. If the surface potential is V_0 volts, one merely has to multiply the results shown here by the factor V_0. In these and the following figures R represents the radial distance from the z-axis. Obviously, the charge densities approach infinity near the edges, as would be expected. This is due to our assumption that edges are infinitely sharp. Also it is clear that over most parts of both end and side surfaces the charge densities are fairly uniform. Only near the edges do the charge densities change rapidly. This behavior allows one to treat most of the surface areas as uniformly charged surfaces. One would therefore expect that the motion of a small particle (small compared to the dimension of the surface) near the center part of an end surface will be similar to that near a charged infinite plane, while the motion near the equator of the cylindrical surface will be similar to that near a charged infinitely long cylinder.

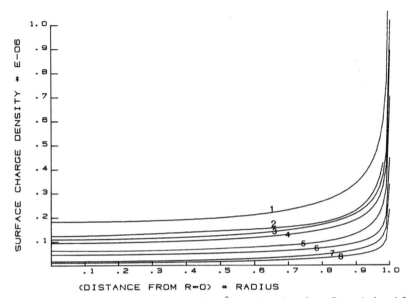

FIG. 4.2. Charge density distributions (in C m^{-2}) on the end surface. Curve 1, $b = 1.21$; curve 2, $b = 1.43$; curve 3, $b = 1.54$; curve 4, $b = 1.67$; curve 5, $b = 2.22$; curve 6, $b = 3.33$; curve 7, $b = 5.00$; curve 8, $b = 8.33$.

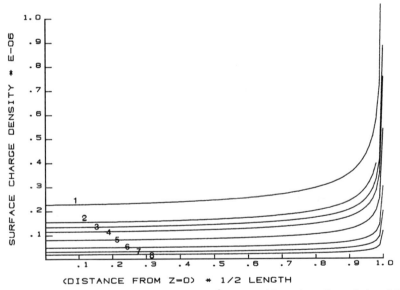

FIG. 4.3. Charge density distributions (in C m^{-2}) on the side surface. Curve 1, $b = 1.21$; curve 2, $b = 1.43$; curve 3, $b = 1.54$; curve 4, $b = 1.67$; curve 5, $b = 2.22$; curve 6, $b = 3.33$; curve 7, $b = 5.00$; curve 8, $b = 8.33$.

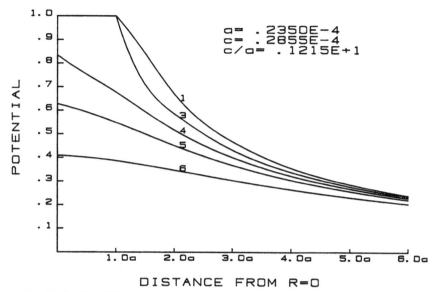

FIG. 4.4. Profiles of the electric potentials for $b = 1.21$. Curve 1, $z = 0$; curve 2, $z = 0.5c$; curve 3, $z = 1.0c$; curve 4, $z = 1.5c$; curve 5, $z = 2.0c$; curve 6, $z = 3.0c$.

As mentioned previously, the ice crystal growth rate (dm/dt, where m is the crystal mass) is analogous to total charge; thus the charge density distributions shown in Figures 4.2 and 4.3 can be viewed as the "growth rate per unit area," which is really the vapor flux density. Therefore, from these two figures, we can also deduce that vapor flux density is the highest at the corner of the column.

Figures 4.4 and 4.5 show the distribution of potential fields (in volts). Curves 1 to 3 represent the potentials at $z = 0$, $0.5c$, and $1.0c$, respectively. In Figure 4.5, curve 2 is often very close to curve 1 and therefore is not shown. As expected, the potentials fall off with increasing distance from the cylinder, the rates of fall (dV/dr, where r is measured in units of a) being larger for shorter cylinders. Also, the potentials converge to one value when far away from the cylinder, regardless of z. This is, of course, because any finite cylinder, when viewed at sufficient distance, will look like a point charge, and therefore the equipotential surfaces approach concentric spheres. The convergence is faster for shorter cylinder because one need not go too far to view them as point charges. One can also view their shapes as being closer to spherical. These two figures also represent vapor density distributions because, as previously noted, vapor density is analogous to electric potential.

Figures 4.6 and 4.7 show the magnitudes of the electric fields (in V/m). There is also a convergence feature which is similar to that of potentials and can be explained in the same manner. Curve 3 is the field at $z = 1.0c$, its value at $R = 1.0a$ being

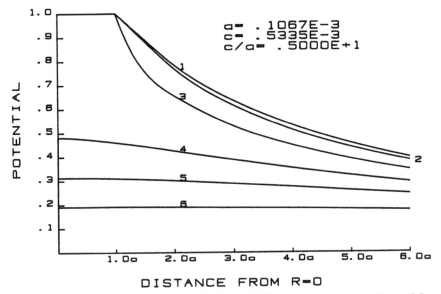

FIG. 4.5. Profiles of the electric potentials for $b = 5.00$. Curve 1, $z = 0$; curve 2, $z = 0.5c$; curve 3, $z = 1.0c$; curve 4, $z = 1.5c$; curve 5, $z = 2.0c$; curve 6, $z = 3.0c$.

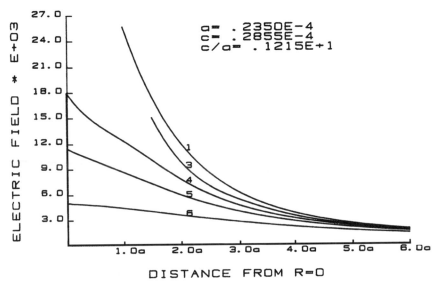

FIG. 4.6. Profiles of the magnitudes of the electric fields for $b = 1.21$. Curve 1, $z = 0$; curve 2, $z = 0.5c$; curve 3, $z = 1.0c$; curve 4, $z = 1.5c$; curve 5, $z = 2.0c$; curve 6, $z = 3.0c$.

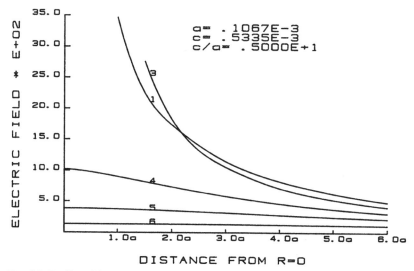

FIG. 4.7. Profiles of the magnitudes of the electric fields for $b = 5.00$. Curve 1, $z = 0$; curve 2, $z = 0.5c$; curve 3, $z = 1.0c$; curve 4, $z = 1.5c$; curve 5, $z = 2.0c$; curve 6, $z = 3.0c$.

infinite. This is, of course, due to the sharp edges in our assumption. Figures 4.8 and 4.9 show the zenith angles of the electric fields at different positions, where the zenith angle is defined as the angle between the electric field vector and the z-axis. If the field vector is pointing straight upward (i.e., along the z-axis), then the zenith angle is zero. Since the electric field is azimuthally symmetric, this angle alone is sufficient to determine the direction of the electric field. These angles, together with the magnitudes given in Figures 4.6 and 4.7, completely determine the electric field vector. Figure 4.10 shows an example of the patterns of the field lines. Such configurations make it clear that the very strong fields near the sharp edges will have an important influence on the dynamic behavior of small charged particles approaching them.

Since prolate spheroids have been considered good approximations of columnar ice crystals previously, it is instructive to compare their field distributions with the present cases. Figure 4.11 compares the potential distributions of circular cylinders and prolate spheroids of analogous dimensions. The semimajor and semiminor axes of the prolate spheroid correspond to the half-length and the radius of the cylinder, respectively. The potential distribution external to a conducting prolate spheroid charged to a surface potential of 1 volt is (Morse and Feshbach, 1953)

$$V = \ln[(\xi + 1)/(\xi - 1)]\big/\ln[(\xi_0 + 1)/(\xi_0 - 1)] \tag{4.18}$$

where $\xi = (r_1 + r_2)/f$ is the generalized radial coordinate in a prolate spheroidal coordinate system; $\xi = $ constant, being a prolate spheroid with interfocal

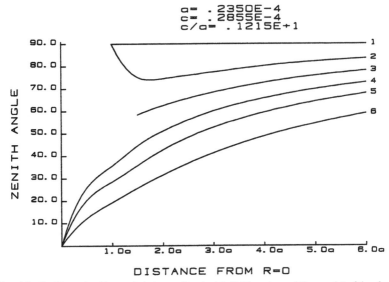

FIG. 4.8. Zenith angles (the angle between the electric field vector and the z-axis) of the electric fields for $b = 1.21$. Curve 1, $z = 0$; curve 2, $z = 0.5c$; curve 3, $z = 1.0c$; curve 4, $z = 1.5c$; curve 5, $z = 2.0c$; curve 6, $z = 3.0c$.

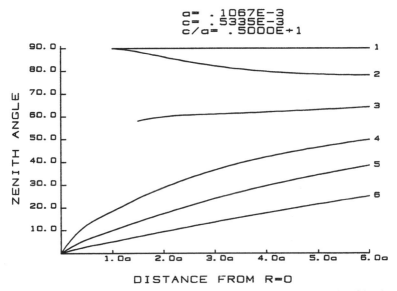

FIG. 4.9. Zenith angles (the angle between the electric field vector and the z-axis) of the electric fields for $b = 5.00$. Curve 1, $z = 0$; curve 2, $z = 0.5c$; curve 3, $z = 1.0c$; curve 4, $z = 1.5c$; curve 5, $z = 2.0c$; curve 6, $z = 3.0c$.

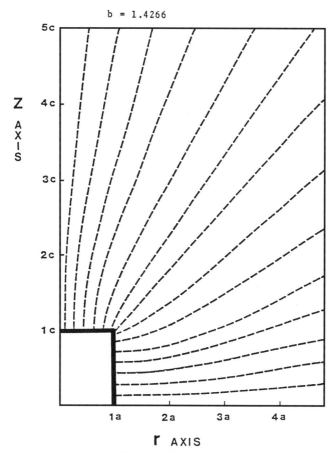

FIG. 4.10. Electric field lines for $b = 1.43$.

distance f; r_1 and r_2 are the distances from an external point P to the two foci; and ξ_0 is the generalized surface of the prolate spheroid, being equal to $2c/a$ where c and a are the semimajor and semiminor axes of the spheroid. It is seen from Figure 4.11 that there are clear discrepancies between the potentials of a prolate spheroid and that of a circular cylinder, both in magnitude and fall-off rate. For example, the potential of the spheroid at $r = 3.0a$ from the equator is $\sim$0.35 V, while that of a circular cylinder is $\sim$0.47 V, a difference of about 34%. This also means that the temperature and vapor density (which are analogous to the electric potential) at this point surrounding a prolate spheroid will also be $\sim$34% lower than for a circular cylinder. In addition, because of the steeper slopes of the potential curves for a prolate spheroid, the electric fields (and hence the heat and vapor fluxes) are overestimated. Since the electric fields

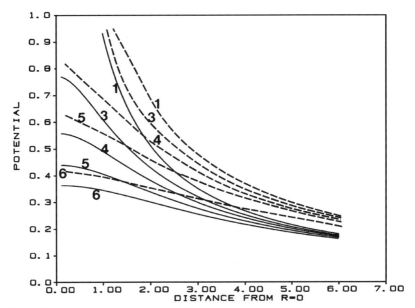

FIG. 4.11. Comparison of the potential distributions surrounding a charged circular cylinder and a charged prolate spheroid. Dashed curves represent a circular cylinder of $b = 1.21$ whereas solid curves represent a prolate spheroid with a and c as the semiminor and semimajor axes, respectively. Zenith angles (the angle between the electric field vector and the z-axis) of the electric fields for $b = 1.21$. Curve 1, $z = 0$; curve 2, $z = 0.5c$; curve 3, $z = 1.0c$; curve 4, $z = 1.5c$; curve 5, $z = 2.0c$; curve 6, $z = 3.0c$.

and the heat and vapor fluxes represent the strengths of electric, thermophoretic, and diffusiophoretic forces, it is clear that using a prolate spheroid to approximate a columnar ice crystal would result in a significant overestimate of these forces.

This, in turn, results in inaccurate estimates of the collision efficiencies of small aerosol particles or droplets with these crystals. These inaccuracies of the prolate spheroidal approximation are in addition to the fact that prolate spheroids lack the two sharp edges that the real columnar crystals possess. In summary, it is felt that the approximation of columnar ice crystals by prolate spheroids is inadequate.

The temperature and vapor density fields surrounding a stationary ice crystal satisfy the same Poisson and Laplace equations as Eq. (4.4) with appropriate boundary conditions

$$T = T_s, \qquad \rho_v = \rho_{v,s} \text{ at the surface}$$
$$T = T_\infty, \qquad \rho_v = \rho_{v,\infty} \text{ at infinity}$$

(4.19)

where T_s, T_∞, $\rho_{v,s}$, and $\rho_{v,\infty}$ are all constants.

Although the equations are the same as for the electric potential case, the boundary conditions are different. The boundary conditions for electric potential are that $V = 1$ at the surface and $V = 0$ at infinity. Therefore, the preceding values of potentials and electric fields cannot be used to represent the respective temperature (vapor density) and heat flux (vapor flux) directly, but must be modified by factors derived in the following equations. Define a dimensionless quantity T' as

$$T' = \frac{T - T_\infty}{T_s - T_\infty} \tag{4.20}$$

Then T' also satisfies Eq. (4.4), with the boundary conditions for T' being

$$\begin{aligned} T' = 1 \quad &\text{at the surface} \\ T' = 0 \quad &\text{at infinity} \end{aligned} \tag{4.21}$$

which are the same as that for V. Therefore, the values given in Fig. 4.4 can be used to represent T'. The actual temperature is, of course,

$$T = (T_s - T_\infty)T' + T_\infty \tag{4.22}$$

and the actual temperature gradient is

$$\nabla T = (T_s - T_\infty)\nabla T' \tag{4.23}$$

Similarly,

$$\rho_v = (\rho_{v,s} - \rho_{v,\infty})\rho_v' + \rho_{v,\infty} \tag{4.24}$$

$$\nabla \rho_v = (\rho_{v,s} - \rho_{v,\infty})\nabla \rho_v' \tag{4.25}$$

where

$$\rho_v' = \frac{\rho_v - \rho_{v,\infty}}{\rho_{v,s} - \rho_{v,\infty}} \tag{4.26}$$

4.2.3. Remarks

In the above, we have presented the mathematical methods and results for the electric, temperature, and vapor density fields surrounding columnar ice crystals. Actually, any quantity satisfying the same Poisson and Laplace equations with Dirichlet boundary conditions (i.e., the values of the dependent variable at the boundaries are specified) can be treated in the same manner. The computed results should be useful in some cloud physical calculations.

Note that we have treated here the cases for stationary ice crystals. The motions of crystals will not influence the electrostatic fields, and the values obtained above can be used directly. The temperature and vapor density fields, however, will change if ice crystals are moving, thus requiring knowledge of the flow fields. The results presented so far can be applied to cases when flow is not important.

In light of what we have developed in Section 2, it may be possible to obtain the various distributions shown here using simple mathematical expressions from Section 2 for the inner boundaries for solving this type of problems.

4.3. Ventilation Coefficients for Falling Ice Crystals

4.3.1. Background

The vapor distributions and other quantities shown in Section 4.2 were obtained without regard to motion. In order to determine the real diffusional growth rates of these particles, it is necessary to recognize that these particles are moving relative to the air. This relative motion causes the air to flow around the crystal in a complicated way which, in turn, influences the distribution of water vapor around the crystal. Since its diffusional growth rate is determined by the vapor density gradient, it is obviously influenced by the motion. This effect on the growth (or evaporation if the air is subsaturated) of ice crystals (and other cloud and precipitation particles in general) is known as the *ventilation effect,* due to which a falling hydrometeor will grow (or evaporate) faster than when it is stationary relative to the air. This is caused by the enhancement of the mean vapor density gradient around the ice crystal. The magnitude of this enhancement is given by a factor called the *mean ventilation coefficient,* defined as (Pruppacher and Klett, 1997)

$$\bar{f}_v = \frac{(dm/dt)}{(dm/dt)_0} \tag{4.27}$$

where the numerator and denominator represent the growth rate of a hydrometeor with mass m under moving and stationary conditions, respectively.

Unlike in the more thoroughly studied cases of water droplets (see Pruppacher and Klett, 1997, Chap. 13, for a review), little information is available for ice crystals. Either the mean ventilation coefficient can be measured experimentally in the laboratory or it can be computed theoretically. Experimental measurements, especially for ice crystals, require sophisticated equipment and are difficult to perform; thus far, only a few direct measurements have been done for ice spheres and hexagonal ice plates (Thorpe and Mason, 1966) and some indirect information has been inferred from experimental studies of snow crystal growth (Takahashi *et al.,* 1991). Theoretical computations of the ventilation coefficients have been carried out by Brenner (1963), who obtained analytical solutions of the convective diffusion equation for infinitely thin circular disks, and by Masliyah and Epstein (1971) and Pitter *et al.* (1974), who computed numerically the ventilation coefficients of thin oblate spheroids of various axis ratios used for approximating hexagonal plates.

The results reported below are obtained using numerical techniques similar to that of Masliyah and Epstein (1971) and Pitter *et al.* (1974), but without using the thin oblate spheroid approximation for hexagonal plates. Instead, the true shapes of hexagonal plates are used directly. In addition, the ventilation coefficients for falling columnar and broad-branch ice crystals are also computed. The three types of crystals considered in the present study are the same as those in Wang and Ji (1997; see Fig. 3.1). The mathematics and physics of the problem, and the numerical methods employed to obtain solutions, are given in the following sections.

4.3.2. Physics and Mathematics

The theoretical problem of determining the ventilation coefficients for falling ice crystals is basically a convection diffusion problem for water vapor around the crystal. Unlike in the stationary crystal case treated in Section 4.2, we will now take the motion of the falling ice crystals into account. The convective diffusion equation appropriate for this situation is

$$\frac{\partial \rho_v}{\partial t} = D_v \nabla^2 \rho_v - \mathbf{V} \cdot \nabla \rho_v \tag{4.28}$$

where ρ_v is vapor density, D_V is the diffusion coefficient of water vapor in air, and $\mathbf{V}$ is the local air velocity vector. The boundary conditions are

$$\rho_v = \rho_{v,s} \quad \text{at the surface of the crystal}$$
$$\rho_v = \rho_{v,\infty} \quad \text{far away from the crystal} \tag{4.29}$$

where $\rho_{v,s}$ and $\rho_{v,\infty}$ are two constants representing the vapor density at the surface of the crystal and far away from it, respectively. For spheres, the inner boundary can be easily written down as the surface where $r = a$ (a is the radius of the sphere); however, the surface of an ice crystal cannot be easily expressed mathematically. The mathematical expressions developed in Section 2 have not been applied to the present study yet, so the boundary conditions are specified numerically.

Equation (4.28) is written in dimensional form. To facilitate the numerical calculations, this equation is nondimensionalized by introducing the following non-dimensional quantities:

$$\phi = \frac{\rho_v - \rho_{v,\infty}}{\rho_{v,s} - \rho_{v,\infty}}$$
$$x' = x/a$$
$$t' = tV_\infty/a \tag{4.30}$$
$$\mathbf{V}' = \mathbf{V}/V_\infty$$
$$N_{Pe} = 2aV_\infty/D_v$$

All the LHS quantities are dimensionless; x, y, and z are the three Cartesian coordinates, t is the time, V_∞ is the free stream velocity (or the air velocity far away from the ice crystal surface), a is the radius of the crystal, and N_{Pe} is the Peclet number. Using the new dimensionless variables, Eq. (4.28) becomes

$$\frac{\partial \phi}{\partial t} = \frac{2}{N_{Pe}} \nabla^2 \phi - \mathbf{V} \cdot \nabla \phi \qquad (4.31)$$

and the boundary conditions (4.29) become

$$\phi = 1 \quad \text{at the surface of the crystal}$$
$$\phi = 0 \quad \text{far away from the crystal} \qquad (4.32)$$

Equations (4.31) and (4.32) constitute the nondimensional set of equations to be solved numerically. While the first (inner) boundary condition can be applied in a straightforward manner, some considerations have to be given before the second (outer) boundary condition can be implemented in the actual computations. The ideal theoretical outer boundary is normally put at $r \to \infty$, which is obviously impossible to do in a real numerical scheme such as that on which the present study is based. Thus some finite outer boundary surfaces have to be devised sufficiently far from the crystal to replace the ideal one.

There are additional considerations for setting the boundary conditions. First, the solution of the convective diffusion equation (4.28) requires knowledge of the local air velocity vector at each point of the numerical grid. These velocity vectors are obtained from the numerical solutions to the time-dependent Navier–Stokes equations for incompressible flow past ice crystals, as reported in Section 3.5. Thus, for consistency of precision, we must use a numerical grid that is either the same as or smaller than the one used in Wang and Ji (1997) for flow field calculations. We choose to use the same grid and hence the same boundary surfaces for the present study. The locations of the outer boundaries are given in Table 3.1.

Second, while the second condition in (4.32) can be applied at the upstream and lateral outer boundaries, it will encounter difficulty at the downstream outer boundary if the flows are unsteady. Owing to the restrictions upon computing resources and hence the distance to the finite downstream boundary, the requirement of a constant ϕ at this distance is most likely unrealistic. Here we replace this condition by the following:

$$\frac{\partial \phi}{\partial z} = 0 \qquad (4.33)$$

This simply means that we require the ϕ field to be continuous at the downstream boundary surface. We have successfully used a similar condition for velocity for solving the flow field problem (Wang and Ji, 1997), and, as it turned out, have had good results for ϕ.

TABLE 4.1 REYNOLDS NUMBERS, DIMENSIONS,
AND CAPACITANCE OF COLUMNAR ICE CRYSTALS IN
THE PRESENT STUDY[a]

N_{Re}	Diameter	Length	Capacitance
0.2	2.0	2.85	1.3628
0.5	2.0	2.85	1.3628
0.7	2.0	3.08	1.4054
1.0	2.0	3.33	1.4535
2.0	2.0	4.44	1.6511
5.0	2.0	6.67	2.0151
10.0	2.0	10.00	2.5067
20.0	2.0	16.67	3.3959

[a] The quantities are dimensionless.

Once the ϕ profile is determined, the growth rate of the falling ice crystal (without considering the coupling of latent heat released or consumed) can be calculated using

$$\frac{dm}{dt} = - \oint_s D_v \nabla \phi \cdot d\mathbf{S} \tag{4.34}$$

where the integration is to be carried out over the ice crystal surface S. On the other hand, the growth rate of a stationary ice crystal is given by the classical electrostatic analog (see Pruppacher and Klett, 1997):

$$\left(\frac{dm}{dt}\right)_0 = -4\pi C D_v (\rho_{v,s} - \rho_{v,\infty}) \tag{4.35}$$

where C is the capacitance of the ice crystal. Hence, in order to calculate the growth rates of stationary ice crystals, it is necessary to know their capacitance.

For columnar ice crystals, which are approximated by finite circular cylinders in this study, we used the formulation of Smythe (1956, 1962) and Wang et al. (1985) to calculate the capacitance. The dimensions and capacitances in this study are given in Table 4.1. For the capacitance of hexagonal plates, we used the formulation of McDonald (1963), who measured the capacitance of various conductors cut in the shape of snow crystals. The theoretical values of C for hexagonal plates of small thickness may be written as

$$C = \frac{1.82a}{\pi} \left(1 + \frac{\Delta S}{S}\right) \tag{4.36}$$

where a is the radius (measured from the center to one of the edges), S is the area of the basal plane, and ΔS is the area difference between the hexagon and a circle with the same radius. The dimensions and the capacitance of the hexagonal plates calculated according to this formula are given in Table 4.2.

TABLE 4.2 REYNOLDS NUMBERS, DIMENSIONS,
AND CAPACITANCE OF HEXAGONAL ICE PLATES
IN THE PRESENT STUDY[a]

N_{Re}	Diameter	Thickness	Capacitance
1.0	2.0	0.2250	0.7298
2.0	2.0	0.1770	0.6977
10.0	2.0	0.1265	0.6639
20.0	2.0	0.1034	0.6485
35.0	2.0	0.0863	0.6371
60.0	2.0	0.0725	0.6278
90.0	2.0	0.0640	0.6221
120.0	2.0	0.0576	0.6179

[a] The quantities are dimensionless.

Unfortunately, there are no accurate values of either measured or theoretically calculated capacitance available for broad-branch crystals. Thus, in this case, we compute the growth rate of stationary broad-branch crystals directly by numerically solving the (nonconvective) diffusion equation first and then use (4.27) to determine the ventilation coefficient. The dimensions of the broad-branch crystals involved in the present study are given in Table 4.3.

When flow is unsteady and eddy shedding occurs, the computed ventilation coefficient will vary slightly with time step size. The final value of the coefficient is taken as the average value over one eddy shedding cycle. Since the eddy shedding occurs mainly downstream, its influence on the coefficient is not very large, typically smaller than 10%.

The numerical scheme (including the grid and the iteration and interpolation techniques) used in the present study is identical to that of Wang and Ji (1997) for obtaining the flow fields around falling crystals.

TABLE 4.3 REYNOLDS NUMBERS AND DIMENSIONS
OF BROAD-BRANCH ICE CRYSTALS IN THE
PRESENT STUDY[a]

N_{Re}	Diameter	Thickness
1.0	2.0	0.15
2.0	2.0	0.14
10.0	2.0	0.0914
20.0	2.0	0.080
35.0	2.0	0.0667
60.0	2.0	0.060
90.0	2.0	0.052
120.0	2.0	0.047

[a] The quantities are dimensionless.

4.3.3. Results and Discussion

As mentioned previously, the ventilation coefficients of three different ice crystal habits of various dimensions (as listed in Tables 4.1, 4.2, and 4.3) are computed as described in Section 4.3.2. The atmospheric pressure is assumed to be $P = 800\,\text{hPa}$ and the temperature $T = -8°\text{C}$. It may seem unnecessary to define P and T for the calculations of the ventilation coefficients, as they are controlled strictly by hydrodynamics. But the pressure and temperature do affect the values of several nondimensional characteristic numbers to be introduced below, and therefore need to be specified.

Figure 4.12 shows an example of the vapor density field around a stationary columnar ice crystal. The field obviously possesses symmetry with respect to the crystal since no motion is involved here. Once motion is introduced, however, the symmetry disappears, and the resulting vapor density fields show enhanced gradients upstream and relaxed gradients downstream, as illustrated by Figures 4.13 and 4.14 for columnar crystals falling at Reynolds numbers 2 and 10, respectively. Comparison of these two figures clearly shows that the higher the Reynolds number, the more pronounced the asymmetry of the vapor density fields and the greater the enhancement of the upstream gradients. Figure 4.15 shows the vapor density fields around falling broad-branch crystals at Reynolds number 2. The main features are essentially the same as those in the previous two figures.

For higher Reynolds number cases, where the flow fields become unsteady, the vapor density fields also become unsteady, but the main features of front enhancement and rear relaxation of vapor density fields remain the same.

The mean ventilation coefficient is then calculated using (4.34), (4.35), and (4.27), and the results are summarized in Figure 4.16a. The horizontal axis is a dimensionless number X defined as

$$X = (N_{\text{Sc,v}})^{1/3}(N_{\text{Re}})^{1/2} \tag{4.37}$$

where $N_{\text{Sc,v}}$ is the Schmidt number of water vapor ($=$ air kinematic viscosity/water vapor diffusivity) and N_{Re} is the Reynolds number of the falling ice crystal. Figure 4.16b shows the correspondence between the Reynolds numbers and X if the Schmidt number is assumed to be constant ($= 0.63$). It is seen that the functional dependence of f_v on X is similar to that found by previous investigators (see Pitter *et al.*, 1974, for a summary). The results for hexagonal plates are very close to those obtained by Pitter *et al.* (1974) for thin oblate spheroids of axis ratio 0.05. The largest difference is only about 10%, indicating that the thin oblate spheroid approximation to a hexagonal plate is a fairly good one. The differences are probably due to the different aspect ratios and the slightly different cross-sectional shapes of the crystals. Thorpe and Mason's (1966) results are somewhat larger than those obtained by us and Pitter *et al.*, but are close to Masliyah and Epstein's (1971) numerical results for oblate spheroids of axis ratio 0.2. Apparently, the axis

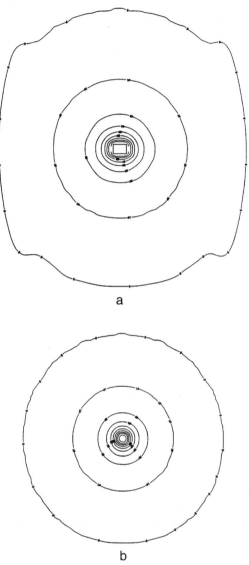

a

b

FIG. 4.12. Water vapor density distribution around a stationary ice column. (a) Length view; (b) end view.

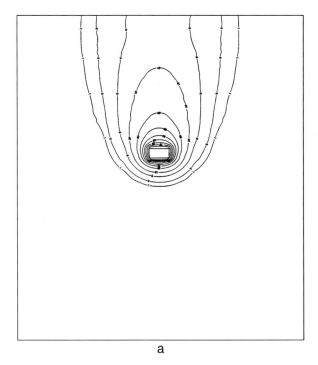

a

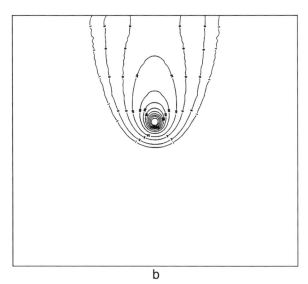

b

FIG. 4.13. Water vapor density distribution around a falling ice column at Re = 2.0. (a) Length view; (b) end view. The contour levels are (from outside) 1, 2, 5, 10, 20, 30, 40, 50, 60, 70, 80, and 100 (surface).

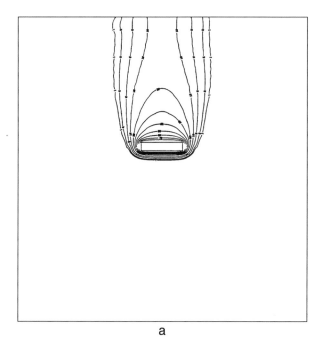

a

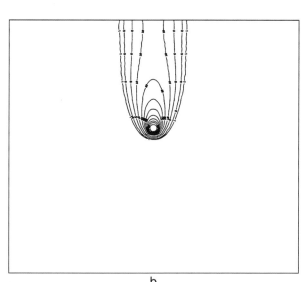

b

FIG. 4.14. Water vapor density distribution around a falling ice column at Re = 10.0. (a) Length view; (b) end view. The contour levels are (from outside) 1, 2, 5, 10, 20, 30, 40, 50, 60, 70, 80, and 100 (surface).

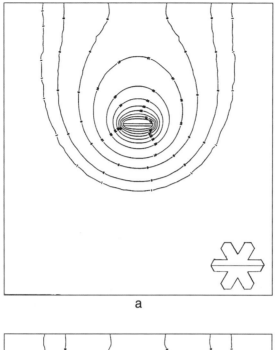

a

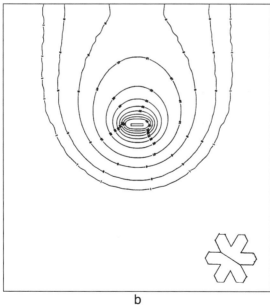

b

FIG. 4.15. Water vapor density distribution around a falling broad-branch crystal at Re = 2.0. (a) Central cross- sectional view; (b) diagonal cross-sectional view. The contour levels are (from outside) 1, 2, 5, 10, 20, 30, 40, 50, 60, 70, 80, and 100 (surface).

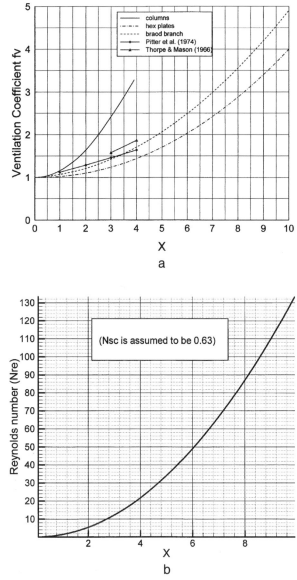

FIG. 4.16. (a) Computed mean ventilation coefficients as a function of the dimensionless parameter X defined in Eq. (4.37). Experimental results of Thorpe and Mason (1960) and numerical results of Pitter *et al.* (1974) are also plotted for comparison. (b) Reynolds number of ice crystals as a function of X assuming $N_{Sc} = 0.63$.

ratio is one parameter that must be considered in characterizing the ventilation coefficient. At present we are aware of no experimental measurements to verify other results.

The three curves in Figure 4.16a can be fitted by the following empirical expressions:

$$\bar{f}_v = 1.0 - 0.00668\,(X/4) + 2.39402\,(X/4)^2 + 0.73409\,(X/4)^3$$
$$- 0.73911\,(X/4)^4 \tag{4.38}$$

for columnar ice crystals of $0.2\,(a = 23.5, L = 67.1\ \mu m) \leq N_{\text{Re}} \leq 20\,(a = 146.4,$ $L = 2440\ \mu m)$;

$$\bar{f}_v = 1.0 - 0.06042\,(X/10) + 2.79820\,(X/10)^2 - 0.31933\,(X/10)^3$$
$$- 0.06247\,(X/10)^4 \tag{4.39}$$

for simple hexagonal plates of $1.0\,(a = 80, h = 18\ \mu m) \leq N_{\text{Re}} \leq 120\,(a = 850,$ $h = 49\ \mu m)$; and

$$\bar{f}_v = 1.0 + 0.35463\,(X/10) + 3.55338\,(X/10)^2 \tag{4.40}$$

for broad-branch crystals of $1.0\,(a = 100, h = 15\ \mu m) \leq N_{\text{Re}} \leq 120\,(a = 1550,$ $h = 73\ \mu m)$.

In the present study, $N_{\text{Sc},v}$ is held constant $(= 0.63)$, so what the figure shows is essentially the variation of the ventilation coefficient with the Reynolds number. Figure 4.17 shows this relation. Here we see that the dependence is nearly linear. However, it is unlikely that the linear trend in the columnar case can be continued much further, and likely that the slope will become smaller at higher Reynolds number.

Figure 4.17 also shows that, at the same Reynolds number, the columnar ice crystal has the highest ventilation coefficient. This is probably because the characteristic dimension of the column used in defining its Reynolds number is its radius instead of its length, so a columnar crystal with a small Reynolds number is actually long and hence has a high fall velocity. This high velocity (higher than for both the hexagonal plate and the broad-branch crystal at the same N_{Re}) is the main reason for its higher ventilation coefficient. It is less obvious why the ventilation coefficient of a broad-branch crystal is higher than a hexagonal plate at the same N_{Re}, but again this may be explained by looking at the dimension of the crystals. Due to the more skeletal structure of the broad-branch crystal, its bulk density is smaller than for a simple hexagonal plate. Hence, at the same Reynolds number, the broad-branch crystal is larger than a hexagonal plate, and the surface area subject to the ventilation effect is correspondingly greater.

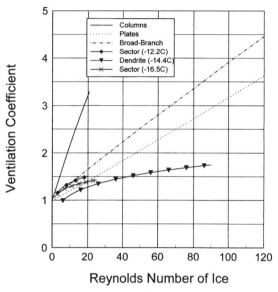

FIG. 4.17. Computed mean ventilation coefficients as a function of the Reynolds number of the falling ice crystal assuming $N_{Sc} = 0.63$.

4.3.4. Remarks

The above ventilation calculations were based on the formulation of water vapor diffusion, which is a mass transfer process. However, the results are also applicable to the ventilation of heat for the same falling ice crystals since the mathematical equations involved (including the boundary conditions) are entirely the same except that the vapor density is replaced by temperature and the vapor diffusivity by the thermal conductivity of air. The reasoning is the same as that stated in Section 4.2.2. In the calculations of the diffusional growth rates of ice crystals where the heat diffusion is coupled with vapor diffusion, it is adequate to set

$$\bar{f}_v = \bar{f}_h$$

where the latter represents the mean ventilation coefficient of heat. Detailed discussion of the equivalence of these two phenomena is given by Pruppacher and Klett (1997).

In the above, we have shown the calculations of ventilation coefficients for falling ice crystals of columnar, hexagonal plate, and broad-branch habits based on the convective diffusion theory of water vapor. The convective part of the mass transfer was computed using the detailed flow fields computed by solving numerically the corresponding Navier–Stokes equations for falling ice crystals. The results show that the ventilation effect can be significant for their diffusional growth. A falling ice crystal may grow several times faster than a stationary one. Similarly, the heating or cooling due to the falling ice crystal's growth or evaporation can also be several times as large as for a stationary one. Clearly, both effects have a significant impact on the evolution of clouds that contain ice crystals, especially for the cirrus clouds, in which nearly all particles are ice crystals.

Note that aside from the ventilation effect examined here, radiative heating/ cooling (near the cloud top and base) may influence the growth rates of ice crystals. These radiative effects and the effects of ambient dynamic conditions on the evolution of cirrus clouds will be examined by a cirrus model in Section 6.

4.4. Collision Efficiencies of Ice Crystals Collecting Supercooled Droplets

4.4.1. Background

We now turn to another important growth mode of ice crystals—their collisional growth. The collision of supercooled cloud droplets with, and the subsequent freezing of, these droplets on ice crystals, known as the riming process, plays a fundamental role in the formation of precipitation-sized hydrometeors in clouds (Pruppacher and Klett, 1997; Cotton and Anthes, 1989; Johnson *et al.*, 1993). A recent numerical study by Johnson *et al.* (1994) indicates that more than 70% of the rain water produced in a midlatitude deep convective storm comes from the melting of graupel and hail. Even in subtropical thunderstorms, the melting of graupel and hail accounts for about 50% of the rainwater production (Lin and Wang, 1997). Since graupel and hail are themselves products of the riming process in clouds, it is logical to expect that riming rates have a significant impact on the rain production rates in convective storms.

In addition, riming involves the phase change of water from liquid to solid and hence the release of latent heat into the surrounding air. In a cloud region where riming proceeds rapidly, this heating may become significant enough to influence the thermodynamic structure and, ultimately, the dynamic behavior of the storm.

The riming rate hinges on two quantities: the collision efficiency between ice particles and supercooled droplets, and the coalescence efficiency of the colliding pair. The coalescence efficiency is usually assumed to be 1 (i.e., 100%) since observations indicate that at temperatures lower than $0°C$ a supercooled droplet instantly turns into ice upon collision with an ice surface. The collision efficiency, on the other hand, is a complicated function of ice particle size, shape, density, and velocity as well as droplet size. This subsection addresses the determination of collision efficiencies. Here we shall restrict our discussion to collisions between pristine ice crystals and supercooled droplets. The riming of graupel and hail, where the collectors are usually larger than pristine ice crystals, will not be considered here.

In order to determine the collision efficiency between an ice crystal and a water droplet accurately, one either conducts experimental measurements under a controlled laboratory condition or performs calculations based on rigorous theoretical models. The former is a difficult and often expensive task, and thus far only a few measurements have been done over limited ranges of ice particle size (Sasyo, 1971; Sasyo and Tokuue, 1973; Kajikawa, 1974). On the other hand, theoretical calculations can be relatively economical to perform in comparison with experimental measurements, if properly done. The main limitations are the accuracies of the formulation and computing algorithm and the adequacy of the computer resources, but these can be overcome with reasonable efforts. The present study is based on theoretical calculations.

Some other investigators have performed theoretical calculations of collision efficiencies between ice crystals and cloud droplets. Ono (1969) and Wilkins and Auer (1970) calculated the collision efficiencies between ice disks and droplets based on inviscid flow fields past disks. Pitter and Pruppacher (1974) and Martin *et al.* (1981) calculated the collision efficiency between ice plates and supercooled droplets assuming that the flow past hexagonal plates can be approximated by that past thin oblate spheroids. Schlamp *et al.* (1975) calculated the collision efficiencies between columnar ice crystals and supercooled droplets, assuming that the flow past an ice column can be approximated by that past an infinitely long cylinder. While these studies contributed significantly to our early understanding of the onset of riming, they left room for improvement. Furthermore, all of them assumed that flow fields are steady, but that is not valid for larger ice crystals that fall in unsteady attitude (Pruppacher and Klett, 1997).

As reported in Section 3, Ji and Wang (1989, 1990) and Wang and Ji (1997) calculated the flow fields past three different shapes of ice crystals: hexagonal ice plates, broad-branch crystals, and ice columns. The crystal shapes used in their calculations were more realistic than those mentioned before. Also, unsteady features such as eddy shedding were included in the calculations. These improvements ultimately led to more accurate computation of flow fields. The present study uses the flow fields as determined by Wang and Ji (1997). Using these fields, we calculated the collision efficiencies between ice crystals of the above three shapes and supercooled droplets. The details of formulations, results and conclusions are given below.

4.4.2. *Physics and Mathematics*

The theoretical problem of determining the collision efficiency between an ice crystal and a supercooled cloud droplet mainly involves the solution of the equation of motion for the droplet in the vicinity of the falling ice crystal. Since the motions occur in a viscous medium, namely air, the effect of flow fields must be considered. The flow fields around falling ice crystals are complicated, and are normally obtained by solving relevant Navier–Stokes equations governing the flow. These flow fields are fed into the equation of motion, and the latter is solved (usually by numerical techniques) to determine the "critical trajectory," i.e., the trajectory of the droplet that makes a grazing collision with the crystal (see, for example, Chap. 14 of Pruppacher and Klett, 1997, for an explanation of the grazing trajectory). Finally, the collision efficiency is calculated based on the knowledge of the grazing trajectory.

As indicated above, the first step of determining the collision efficiency is to determine the flow fields around falling ice crystals. This is done by solving the incompressible Navier–Stokes equations for flow past ice crystals as described in Section 3.

The equation of motion for a cloud droplet of radius a_2 in the vicinity of a falling ice crystal of characteristic dimension a_1 is

$$m\frac{d\mathbf{V}}{dt} = m\frac{d^2\mathbf{r}}{dt^2} = \mathbf{F}_g + \mathbf{F}_D \qquad (4.41)$$

where m is the mass of the droplet, $\mathbf{V}$ is its velocity, $\mathbf{r}$ is its position vector, $\mathbf{F}_g$ is the buoyancy-adjusted gravitational force, and $\mathbf{F}_D$ the hydrodynamic drag force due to the flow. These two forces are expressed as

$$\mathbf{F}_g = mg\left(\frac{\rho_w - \rho_a}{\rho_w}\right) \qquad (4.42)$$

where ρ_w and ρ_a are the density of water and air respectively, and

$$\mathbf{F}_D = 6\pi\eta a_2\left(\frac{C_D\text{Re}}{24}\right)(\mathbf{V} - \mathbf{u}) \qquad (4.43)$$

In order to calculate the drag force (4.43), we need to input the local flow velocity $\mathbf{u}$ at each time step. That value comes from the solution of the Navier–Stokes equations.

In order to solve (4.41), it is also necessary to specify an appropriate initial condition, which in this case is the initial horizontal offset, y, of the droplet from the vertical line passing through the center of the falling ice crystal (see Pruppacher and Klett, 1997, p. 569). Needless to say, this offset has to be set at a distance sufficiently upstream to ensure that the droplet is progressing in a straight line at the time. In this study the initial offset was set at 20 radii upstream of the ice crystal, and this proved adequate for the purpose stated above. With this initial condition in place, Eq. (4.41) can be solved for V and hence the drop radius r as a function of time, thus defining its trajectory.

To determine the collision efficiency, we need to determine the critical initial offset y_c of the droplet such that it will make a grazing collision with the ice crystal.

An initial offset greater than y_c would result in a miss, whereas one smaller than y_c would result in a hit. In the present study, a bisection technique similar to that used in Miller and Wang (1989) was used to determine y_c. Once this is done, the next step is determining the collision efficiency E.

4.4.3. Definition of Collision Efficiency for Nonspherical Collectors

Since the collector is an ice crystal that is usually not a sphere, it is important to make a closer examination of the proper definition of collection efficiency here. The old definition of collision efficiency is based on spherical symmetry [e.g., Pruppacher and Klett, 1997, Eq. (14.1)] and is therefore inappropriate here. In the following we shall take a closer look at this definition problem.

Figure 4.18 illustrates the definition of collision efficiency between two spherical water drops. The most important quantity involved is y_c, the initial horizontal offset of the center of sphere a_2 from the vertical through the collector sphere a_1, both identified by their radii such that a grazing collision results (Mason, 1971; Pruppacher and Klett, 1978). The collision efficiency E is then defined as

$$E \equiv y_c^2/(a_1 + a_2)^2 \tag{4.44}$$

(The *linear* collision efficiency is defined as y_c/a_1 but is less popular.) The definition given by (4.44) is essentially two-dimensional and will therefore work for all cases whose configurations are two-dimensional. For example, the collision between two spheres as illustrated in Figure 4.18 is two-dimensional since y_c is independent of the azimuthal angle ϕ. Another example is the collision between an infinitely long cylinder and small cloud droplets; the flow field is again two-dimensional.

But this same definition fails to apply when the configuration is three-dimensional. One example is illustrated in Figure 4.19, where a cylindrical ice crystal of length L collides with small spherical droplets or aerosol particles. Here we see that the value of $y_{c,1}$ will generally differ from that of $y_{c,2}$, because the collector and the flow field around it do not possess rotational symmetry about the fall direction. In this situation, E defined in Eq. (4.43) results in ambiguity. The same difficulty exists for all cases where y_c depends on the azimuthal angle.

Instead of (4.44), the following more general definition of collision efficiency is more suitable for the 3-D situations:

$$E = K/K^* \tag{4.45}$$

where

K = the effective volume swept out by the collector per unit time

K^* = the geometric volume swept out by the collector per unit time

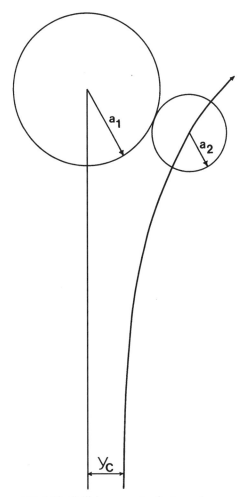

FIG. 4.18. Collision geometry for two spheres.

In Figure 4.19, K is the volume containing the droplets that will be collected by the cylinder per unit time. K^* is simply the geometric volume per unit time, $2aL(V_{\infty,i} - V_{\infty,w})$, where a is the radius and L is the length of the ice crystal, and $V_{\infty,i}$ and $V_{\infty,w}$ are the terminal velocities of ice crystal and water droplet, respectively. In the case of spherical drops, this definition reduces to Eq. (4.44).

The new definition of collision efficiency does not necessarily invalidate previous results. As noted above, the new definition is the same as the old one for particles with rotational symmetry, such as spheres and circular plates. For the special case

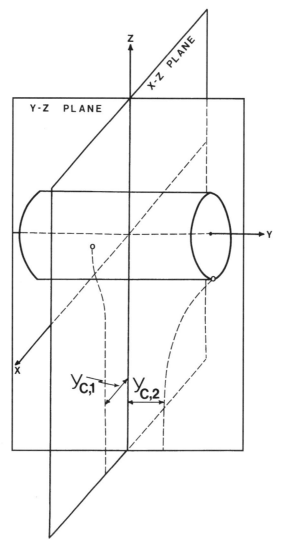

FIG. 4.19. Collision geometry for a finite cylinder and a small sphere.

of an infinitely long cylinder colliding with small droplets, the old definition is still valid (Schlamp *et al.*, 1975) since the flow field is two-dimensional. However, because the cylinder is finite and the geometry is not rotationally symmetric, the new definition must be used.

In the case of the unsteady flow, the trajectory of a droplet starting from a certain initial offset was determined by averaging a few trajectories over an eddy shedding cycle. This was done for a few cases, but was later found to be unnecessary because these trajectories vary very little owing to the fact that grazing collisions in this study all occur in the upstream regions, where flow fields are steady. In addition, droplets are massive enough to resist small fluctuations in the flow fields. This may not be the case if the collection of submicron particles by hydrometeors is considered, since rear capture may occur in that case (e.g., Wang *et al.*, 1978; Wang and Jaroszczyk, 1991) and the unsteady fields in the downstream would have greater effect.

4.4.4. Ice Crystal Collision Efficiencies for Ice Columns, Hexagonal Plates, and Broad-Branch Crystals

Using the formulations described above, we can determine the collision efficiencies of ice crystals colliding with supercooled water droplets. Some of the computational results by Wang and Ji (1999) will be discussed below.

Ice crystal collectors of three different habits are considered below: columnar ice crystals (approximated as finite circular cylinders), hexagonal ice plates, and broad-branch ice crystals. These are the same ice crystals whose dimensions are shown in Tables 4.1, 4.2, and 4.3. Figure 4.20 shows several trajectories of a droplet 2 μm in radius moving around a falling broad-branch crystal of Re = 10. Of the eight trajectories shown here, trajectories 1, 2, 6, 7, and 8 are misses whereas trajectories 3, 4, and 5 are hits. Trajectory 4 is the central trajectory while 3 and 5 are grazing trajectories.

Note that since the collector ice crystals are not spheres, the critical initial offset y_c does not possess circular symmetry, but rather is a function of the azimuthal angle. The asymmetry is most easily shown by the shape of the collision cross section A swept out by radial lines of length $y_c(\theta)$ for all azimuth angles θ. Figures 4.21, 4.22, and 4.23 show examples of these collision cross sections for droplets of various sizes colliding with three types of ice crystals. It is immediately clear that the shapes of the cross sections are more or less similar to those of the ice crystal cross sections themselves. When droplets are small, their collision cross sections (and hence the collision efficiencies) are usually (but not always) smaller. As the droplets become larger, the collisional cross sections become larger and the shapes are closer to the cross sections of the ice crystals. This behavior is obviously due to the inertia of the droplet relative to the strength of the hydrodynamic drag force, reasoning discussed in great detail in Pruppacher and Klett (1997). When droplets are small, their inertias are small compared to the drag and their trajectories are close to the streamlines of the flow fields, which are generally curved around the crystal. Thus the shape of the collision cross section would differ significantly from that of the crystal. When droplets are larger, their inertia becomes greater and their trajectories are straighter, so that the collision cross sections have shapes closer to that of the crystal.

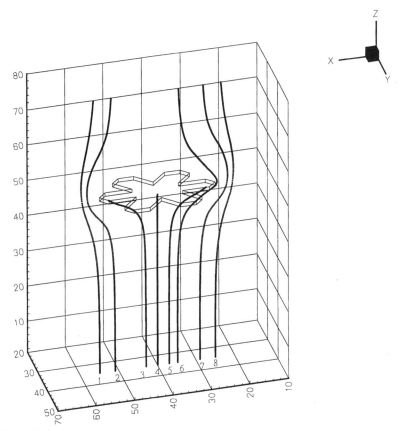

FIG. 4.20. Trajectories of a droplet of 2 μm in radius moving in the vicinity of a falling broad-branch crystal at Re = 10. Trajectories 1, 2, 6, 7, and 8 are misses, and trajectories 3, 4 and 5 are hits.

Hexagonal Plates

Figures 4.24, 4.25, and 4.26 show the computed collision efficiencies for the three crystal habits. Figure 4.24 is for the case of hexagonal ice plates. The general feature here is that, at a fixed crystal Reynolds number, the efficiency of a very small droplet is very small owing to its small inertia, as explained previously. For Re = 1.0 and 2.0, the efficiency drops to very small value ($<10^{-4}$) for droplets with radii <9 μm. For higher Re (larger ice crystals) cases, this efficiency drop is more gradual and there is no sharp cutoff. This is in contrast with earlier studies where a cutoff at $a_2 \approx 5$ μm occurs (e.g., Pitter and Pruppacher, 1974; Pitter, 1977). Instead, the efficiency remains finite even for droplets as small as 2.5 μm, in good agreement with Kajikawa's (1974) experimental results. Recent observational studies also

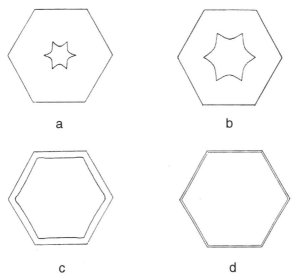

FIG. 4.21. Shape of collision cross sections for a hexagonal ice plate at Re = 20, colliding with supercooled droplets of radius r. The fixed hexagon is the cross section of the ice plate: (a) $r = 3 \mu$m, (b) $r = 5 \mu$m, (c) $r = 11 \mu$m, and (d) $r = 27 \mu$m.

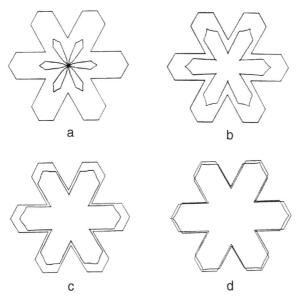

FIG. 4.22. Shape of collision cross sections for a broad-branch crystal at Re = 35, colliding with supercooled droplets of radius r. The fixed hexagon is the cross section of the ice plate: (a) $r = 5 \mu$m, (b) $r = 9 \mu$m, (c) $r = 15 \mu$m, and (d) $r = 36 \mu$m.

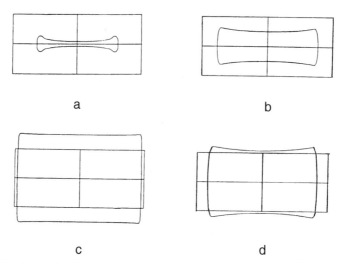

FIG. 4.23. Shape of collision cross sections for a columnar ice crystal at Re = 2.0, colliding with supercooled droplets of radius r. The fixed hexagon is the cross section of the ice plate: (a) $r = 4$ μm, (b) $r = 6$ μm, (c) $r = 35$ μm, and (d) $r = 43$ μm.

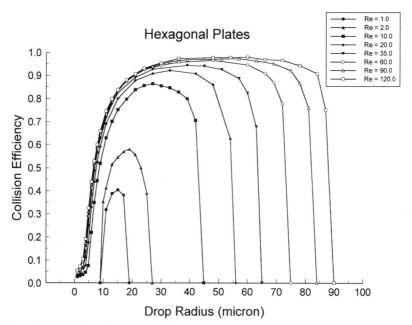

FIG. 4.24. Collision efficiencies of hexagonal ice plates colliding with supercooled water droplets. The last data points (at large drop size end) for Re = 0 to 120 are extrapolated.

confirm that many frozen droplets on the rimed ice crystals are smaller than 5 μm. The scarcity of frozen droplets with radius <5 μm on planar ice crystals in some previous field observations (e.g., Harimaya, 1975; Wilkins and Auer, 1970; Kikuchi and Ueda, 1979; and D'Enrico and Auer, 1978) was probably due to the local microstructure of clouds instead of the intrinsic collision mechanism (Pruppacher and Klett, 1997).

As the drop size increases, the efficiency increases rapidly. The efficiency reaches a peak or plateau, depending on the Reynolds number of the ice crystal, and then drops off sharply for further increasing drop size. The dropoff of efficiency is apparently due to the increasing terminal velocity of the droplet. When the collector ice crystal and the droplet have about the same velocity, collision is nearly impossible and the efficiency becomes very small (Pitter and Pruppacher, 1974; Pitter, 1977; Martin *et al.*, 1981; Pruppacher and Klett, 1997). The efficiency maxima take the shape of peaks in smaller Re cases, but become broader plateaus as the ice crystal Re increases. Owing to their low terminal fall speeds, smaller crystals are quickly "outrun" by the largest droplets, thus preventing a collision.

Broad-Branch Crystals

Figure 4.25 shows the collision efficiencies for broad-branch crystals colliding with supercooled droplets. The main features are similar to those for hexagonal

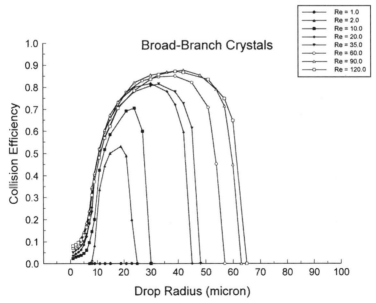

FIG. 4.25. Collision efficiencies of broad-branch crystals colliding with supercooled water droplets. The last data points (at large drop size end) for Re = 0 to 120 are extrapolated.

plates. The collision efficiencies for Re = 1.0 are practically zero, representing an inability to rime. This cutoff of riming ability will be discussed later.

The collision efficiencies of broad-branch crystals are, in general, smaller than for hexagonal plates at the same Reynolds number. The maximum efficiencies in the plateau region are about 0.9, unlike the case of hexagonal plates whose maximum efficiencies are near 1.0. This is probably due to the more open structure of a broad-branch crystal, which would allow the droplet to "slip through" the gap between branches. The plateau is also much narrower than for a hexagonal plate of the same Reynolds number. This is most likely due to the lower fall velocity of the broad branch crystal (as compared to a hexagonal plate at the same Reynolds number), which is thus more easily outrun by a large droplet.

If the above reasoning holds true, it implies that stellar crystals, whose structures are even more open, probably have collision efficiencies similar to or smaller than those of broad-branch crystals. However, the same cannot be said for dendrites, since they have more intricate small branches that enable them to capture droplets with higher efficiency.

Columnar Ice Crystals

Collision efficiencies of columnar ice crystals colliding with supercooled droplets are shown in Figure 4.26. The general features here look very similar

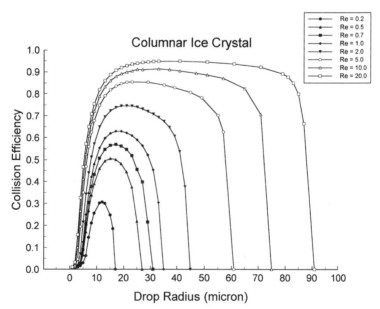

FIG. 4.26. Collision efficiencies of columnar crystals colliding with supercooled water droplets. The last data points (at large drop size end) for Re = 0 to 120 are extrapolated.

to those in Figure 4.24, despite the difference in Reynolds number ranges between these two cases. This is because the difference in Re is a superficial one, since the Reynolds numbers of the falling columns are determined based on their radii instead of lengths. Had the latter been used, the two sets of Reynolds numbers would be much closer in magnitude.

The plateaus in Figure 4.26 are not as flat as those in Figure 4.24, but slope down toward larger drop size. Although the exact cause is not known, this is likely due to the higher asymmetry of a column relative to a plate. The effect of this symmetry would become more pronounced as the drop size increases.

Finite versus Infinite Cylinders

It is educational to examine the differences between the collection efficiencies that result for finite versus infinite cylindrical approximations to columnar ice crystals, so that we can assess the validity of the infinite-cylinder approach used in several previous studies (e.g., Schlamp *et al.,* 1975). Figure 4.27 shows two sets of collection efficiency curves, one for finite cylinders and the other for infinite ones, for Re = 0.5, 1.0, 5.0, and 20.00. For Re = 0.5 and 1.0, we see that the finite cylinders have higher efficiencies than the infinite ones. The difference is greater for Re = 0.5 and becomes smaller for Re = 1.0. For Re = 5.0 and 20.0,

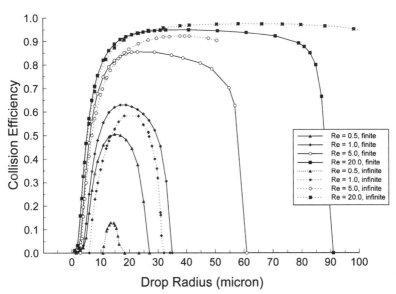

FIG. 4.27. Comparison between the collision efficiencies for finite cylinders (present results) and infinite cylinders (Schlamp *et al.,* 1975) for Re = 0.5, 1.0, 5, and 20.

the two sets of collection efficiencies are almost identical for drop size smaller than 30 μm, becoming more pronounced for larger drops. The enhanced efficiencies of the infinite cylinders may be due to the artificial effect of the superposition technique used by Schlamp et al. (1975), and hence probably do not reflect the real effect of the infinite length assumption.

The above paragraph says that the discrepancies between the two sets of curves are most important when the columns are small and become less significant when column size (and hence the Reynolds number) increases. This is to be anticipated because, in the present study, smaller columns have aspect ratios much different from that of infinitely long cylinders, so that the collision efficiencies would also show greater differences. For larger columns, the aspect ratios are closer to that of infinite cylinders, so that the collision efficiencies are also closer to those of the latter.

In short, using the infinite cylinder approximation in treating the collisions between columnar ice crystals and supercooled drops is valid when the drop size is between a few and about 30 microns and the ice column Reynolds number is greater than 5. Since the flow field around an infinitely long cylinder is easier to compute than that around finite cylinders, this approximation may be useful when computing resources are of concern.

When the drop size becomes smaller than a few microns, the infinite cylinder model underpredicts the collision efficiencies for the same reason discussed in the paragraph about hexagonal plate results.

Riming Cutoff

Earlier observational studies have suggested the existence of a cutoff size (defined as the maximum dimension) of ice crystals below which riming cannot occur (Ono, 1969; Wilkins and Auer, 1970; Harimaya, 1975; Takahashi et al., 1991). Since riming is due to the collision between ice crystals and supercooled droplets, the cutoff would occur at a crystal size where the collision efficiency is zero for all droplet sizes. Earlier theoretical studies of Pitter and Pruppacher (1974) and Pitter (1977) put the riming cutoff size of planar ice crystals at 300 μm, which seemed to agree with observations at the time. However, recent studies indicate riming cutoff sizes smaller than this value (Devulapalli and Collett, 1994). The results shown in Figures 4.24 through 4.26 can be used to predict the cutoff riming crystal size. This is done by plotting the maximum collision efficiency (the peak value of each curve in Figures 4.24, 4.25, and 4.26) as a function of the corresponding crystal size for each crystal habit, as shown in Figure 4.28. The point where the extrapolated curve intersects the x-axis (where $E = 0$) indicates the cutoff crystal size. Using this method, we determined that the riming cutoff size is about 35 μm for columnar ice crystals, 110 μm for hexagonal plates, and 200 μm for broad-branch

Threshold Riming Size

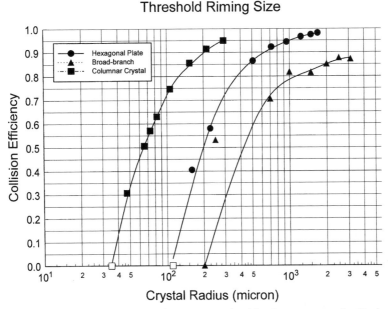

FIG. 4.28. Cutoff riming ice crystal sizes as extrapolated by the present results. For broad-branch crystals, the data point for crystal radius 2.5 μm was ignored when performing the best fit.

crystals. These values are reasonably close to the observations of Wilkins and Auer (1970), Reinking (1979), and Bruntjes *et al.* (1987) as summarized in Table 4.4.

The collision efficiencies of three types of ice crystals colliding with supercooled water drops have been computed and presented above. The main improvements in the present study over previous studies are as follows:

(1) More realistic ice crystal shapes are adopted, especially the finite lengths of the columns and the broad-branch crystals, whose efficiencies have never been reported before.

TABLE 4.4 OBSERVED CRITICAL RIMING SIZE[a]

Crystal habit	Wilkins and Auer (1970)	Reinking (1979)	Bruntjes *et al.* (1987)
Hexagonal plate (Pla)	—	—	$d = 150\ \mu$m
Broad-branch crystal (Plc)	—	$d = 275\ \mu$m	$d = 240\ \mu$m
Columnar crystal (Cle)	$l = 100\ \mu$m, $d = 40\ \mu$m	—	$l = 125\ \mu$m, $d = 40\ \mu$m
Long solid column (Nle)	$l = 100\ \mu$m, $d = 30\ \mu$m	—	—

[a] The code after the crystal habit is Magono and Lee's (1966) classification of natural snow crystals.

(2) More accurate flow fields, including the 3-D and unsteady features, are used to determine the grazing trajectories.

The improvements have been demonstrated by the more accurate prediction of the captured droplet sizes and the cutoff riming crystal sizes. We believe the improved collision efficiency values will lead to more accurate growth rate calculations of ice particles in clouds.

The results presented here pertain to relatively small and pristine ice crystals colliding with small supercooled droplets, so they are mainly applicable to the initial stages of riming process when ice particles have not grown too much. When riming is further along, the ice crystal gradually loses its pristine shape. However, as long as the basic shape of the ice crystal in question is discernible, we believe the present results are still useful for estimating the riming efficiency, as the flow fields would not differ much. As riming goes still further, the original shape of the ice crystals eventually becomes lost, and the pristine ice assumption no longer applies.

As indicated before, the coalescence of the supercooled drop with the ice surface is assumed to be 100%, so that the case where droplets may bounce off from the ice surface is not considered. It is a much more complicated task to determine theoretically the riming rates of larger ice particles, such as graupel and hail, which may fall in zigzag attitude.

5. Scavenging and Transportation of Aerosol Particles by Ice Crystals in Clouds

5.1. Importance of Aerosol Particles in the Atmosphere

Although considered as a variable constituent and present in trace amounts, aerosol particles play many important roles in our atmosphere. First, without aerosol particles there would be little chance of clouds, and hence precipitation, because homogeneous nucleation of water drops is very unlikely to occur in the environment of Earth's atmosphere. Instead, heterogeneous nucleation is necessary for cloud droplets to form, and that requires the presence of cloud condensation nuclei (CCN) which are, of course, aerosol particles. Thus, aerosol particles can be regarded as the initiators of clouds, and it is entirely logical to expect that the amount of aerosol particles in the atmosphere will be correlated with the amount of clouds globally.

But the amount of clouds directly controls how much solar radiation can reach the Earth's surface, and the latter is the central factor in the global climate process. Therefore, the above aerosol–cloud relation implies that the concentration (and the chemical composition) of the aerosol particles has great impact on the global climate. In addition, as is becoming increasingly clear, the amount of aerosol particles produced by human industrial activities may be significant enough to affect solar radiation directly, even without considering their cloud-forming capability.

Aside from this impact on climate, aerosol particles have long been studied for their possible (and mostly adverse) impact on the air quality. The most well-known example is probably the formation of photochemical smog in large cities, such as Los Angeles. The impact on air quality is directly related to the concentration of aerosol particles. For summaries of these impacts, the reader is referred to textbooks such as those of Pruppacher and Klett (1997) and Seinfeld and Pandis (1997).

But aerosol particles also get removed from the atmosphere and the most efficient natural removal mechanism is their capture by cloud and precipitation. Once the particles become attached to a hydrometeor, they can fall out of the atmosphere along with the precipitating hydrometeors. This process is often called precipitation scavenging, which should be considered as a self-cleansing mechanism of the atmosphere. At present, our knowledge of its efficiency is still rather qualitative. But quantifying the efficiency is important in view of the impacts of aerosol particle concentration on global climate and air quality, and therefore it warrants more careful study. This is the main subject of this section.

5.2. Physical Mechanisms of Precipitation Scavenging

There are many physical mechanisms that can result in the removal of aerosol particles by cloud and precipitation particles. In this section, we briefly review the few mechanisms that are thought to be especially important to precipitation scavenging, i.e., the removal of aerosol particles by precipitation.

(1) Nucleation

As mentioned earlier, some aerosol particles, especially hygroscopic ones, can serve as CCN to initiate cloud droplets. Another group of aerosol particles, especially the hydrophobic kind, can serve as the nuclei to initiate ice crystals, and are thus are known as ice nuclei (IN). Upon the occurrence of nucleation, both CCN and IN would become part of the cloud droplet or ice crystal, and hence may fall out with these hydrometeors if the latter grow to precipitation size. For this reason, nucleation is regarded as one of the precipitation scavenging mechanisms and is often called *nucleation scavenging*.

(2) Inertial Impaction

Owing to their different speeds in clouds, hydrometeors may collide with and hence capture aerosol particles, thereby removing them from the atmosphere.

This is called *inertial impaction,* a process analogous to the collision between cloud droplets and ice crystals.

(3) Brownian Diffusion

Small aerosol particles, especially those smaller than 1 μm in radius, move randomly in air because they are constantly bombarded by air molecules in random fashion. This is known as the Brownian motion of small particles. Einstein (1905) studied this problem and showed that the Brownian motion of small particles is kinetically similar to the diffusion of gases, hence the name Brownian diffusion. Due to this motion, aerosol particles may collide with and become captured by hydrometeors, even in the absence of mean relative motion between the two. Evidently, the Brownian effect is more pronounced for smaller particles than for larger ones.

(4) Electrostatic Forces

When both hydrometeors and aerosol particles are electrically charged, electrostatic forces operate between them. If the forces are attractive, the two may collide, resulting in the removal of the aerosol particles. Electrostatic interaction between the two may occur even if only one is electrically charged, due to the so-called "image force"; and even if neither is charged, electrostatic interaction may still occur in the presence of an external electric field that can cause "induced charges" on both particles (see, for example, Lorrain and Carson, 1967; Jackson, 1974).

(5) Phoretic Forces

The phoretic forces of interest here are microscopic scale forces that operate near the surfaces of the collectors. When a hydrometeor is either evaporating or growing, it will consume or release latent heat, respectively. This causes a nonuniformity of temperature, and hence a temperature gradient, near the hydrometeor surface. This temperature gradient in the air generates a microscopic force on the aerosol particles due to the differential thermal agitation of the air molecules. This is called the thermophoretic force. In an evaporating hydrometeor, where the surface is colder than the air, the thermophoretic force points toward the surface (the force is always downgradient), and vice versa for a growing hydrometeor.

But in addition to the temperature gradient, there is also a vapor density gradient near the surface of a growing or evaporating hydrometeor. This vapor density gradient will generate a microscopic flow, called the *Stefan flow,* near the hydrometeor surface, exerting a force on the nearby aerosol particle. This force, called the diffusiophoretic force, points away from the surface of an evaporating hydrometeor because the vapor density gradient points toward the surface; however, the force direction is always downgradient. Again, the case of a growing hydrometeor is just the opposite.

From the descriptions above, it is clear that thermophoretic and diffusiophoretic forces are directed opposite to each other, hence tending to cancel each other. We shall see this later, using more concrete examples. For a more detailed discussion of these phoretic forces, see Pruppacher and Klett (1997, Chap. 17) or Hidy and Brock (1971).

(6) Turbulence

It is commonly accepted that air flow in the atmosphere is generally turbulent and that this turbulence will, conceivably, cause the collection of aerosol particles by hydrometeors and hence their removal from the atmosphere. At present, the scientific community is inclined to believe that turbulence increases the collection efficiency of aerosol particles by hydrometeors, but rigorous proof is still lacking.

In the following sections, we look into the scavenging of aerosol particles by ice particles by mechanisms (2)–(5). This is because the effect of nucleation scavenging is best studied using a cloud model, whereas the approach introduced below utilizes dynamic equations. Similarly, turbulence cannot be treated easily by the dynamic equation approach. Hence we omit mechanisms (1) and (6) and focus on how the remaining four mechanisms affect precipitation scavenging.

5.3. The Theoretical Problem of Ice Scavenging of Aerosol Particles

In this section we investigate the theoretical problem of determining the collection efficiency of aerosol particles by ice crystals. The approach follows that of Wang and Pruppacher (1980b), Martin et al. (1980, 1981), Wang (1985, 1987), Miller and Wang (1989), and Wang and Lin (1995).

The main technique of treating this theoretical problem is similar to that used by Wang et al. (1978) in treating the scavenging of aerosol particles by water drops, namely, a combination of two complementary models to deal with small and large aerosol particles separately.

The model used to treat the scavenging of small aerosol particles (those with radii ≤ 0.5 μm) is called the *flux model*. This model takes into account the combined effect of Brownian diffusion, phoretic forces, and electrostatic forces by explicit dynamic representations, but treats the effect of hydrodynamic drag (hence the inertial impaction) empirically for the following reasons: (a) For such small particles the dominant scavenging mechanism is Brownian diffusion while scavenging due to inertial impaction is small; (b) this model utilizes the convective diffusion equation for determining the flux of aerosol particle concentration; (c) this equation can be solved analytically with the inclusion of phoretic and electrostatic forces but not with hydrodynamic drag force, whose inclusion would make the convective diffusion equation numerically "stiff" and difficult to solve; and (d) inertial impaction, which is most influenced by the drag, is insignificant anyway.

The model used to treat the scavenging of large aerosol particles (those with radii >0.5 μm) is called the *trajectory model*. This model takes into account the combined effect of inertial impaction, phoretic forces, and electrostatic forces, but ignores the effect of Brownian motion. This is because particles in this size range perform little Brownian motion, rendering its effect insignificant. This model is based on the equation of motion for aerosol particles under the influence of external forces (drag, phoretic forces, and electrostatic forces). The solutions of the equation of motion determine the aerosol particle trajectories that result in grazing collisions with the collector. These trajectories determine the collision efficiencies.

The final solutions are the combinations of results from these two models. The joining of the two model results is not done arbitrarily. As will be seen later, there is a broad aerosol size range where the results match each other closely, indicating the true complementary nature of the two models.

5.4. Physics and Mathematics of the Models

We now present the mathematical formulation of the two models described above, first the trajectory model and then the flux model.

5.4.1. The Trajectory Model

This model applies to ice crystals scavenging larger aerosol particles whose radii are between about 0.5 and 10 μm. The precise lower size limit depends on the ambient atmospheric conditions and the crystal size, and may vary between 0.1 and 1.0 μm. The model considers the collection of aerosol particles by inertial impaction, phoretic forces, and electric forces, but neglects Brownian diffusion. The efficiency is computed from an analysis of the trajectories of the aerosol particles moving past the ice crystal. Assuming that the flow around the aerosol particle does not affect the crystal motion (which is usually justified, considering the smallness of the aerosol particles in comparison to the ice crystal), an aerosol particle trajectory can be determined from the equation of motion:

$$m\frac{dv}{dt} = mg^* - \frac{6\pi\eta_a r}{(1 + \alpha N_{Kn})}(v - u) + F_{Th} + F_{Df} + F_e \tag{5.1}$$

Equation (5.1) applies to the motion of an aerosol particle of radius r, mass m, and velocity v moving around an ice crystal of radius a_c, both falling in air of dynamic viscosity η_a under the effect of gravity, plus the hydrodynamic, phoretic, and electric forces. In Eq. (5.1), $g^* = g(\rho_p - \rho_a)/\rho_a$, where g is the acceleration of gravity, ρ_p is the bulk density of the aerosol particle, ρ_a is the density of the air, $N_{Kn} = \lambda_a/r$ is the Knudsen number, λ_a is the free path length of air molecules and, u is the velocity field around the falling collector. Also,

$\alpha = 1.25 + 0.44 \exp(-1.10 N_{Kn}^{-1})$, F_{Th} is the thermophoretic force given by

$$F_{Th} = \frac{12\pi \eta_a r (k_a + 2.5 k_p N_{Kn}) k_a \, \nabla T}{5(1 + 3N_{Kn})(k_p + 2k_a + 5k_p N_{Kn})p} \tag{5.2}$$

and F_{Df} is the diffusiophoretic force given by the relation

$$F_{Df} = -6\pi r \eta_a \frac{0.74 D_{va} M_a \, \nabla \rho_v}{(1 + \alpha N_{Kn}) M_w \rho_a} \tag{5.3}$$

where k_a and k_p are the respective thermal conductivities of air and the aerosol particle, D_{va} is the diffusivity of water vapor in air, T is absolute temperature, p is the air pressure, ρ_v is the water vapor density in air, and M_w and M_a are the respective molecular weights of water and air. Equations (5.2) and (5.3) were evaluated from a knowledge of the vapor density and temperature distributions around the falling crystal. These distributions were determined previously by numerically applying steady-state convective diffusion equation to an ice crystal as described in Section 4.2. The velocity field u around the falling ice crystal is obtained by the method described in Section 3.

The electric force F_e in Eq. (5.1) is assumed to be determined by the electric charges residing on the surfaces of ice crystals and aerosol particles, and by the strength of the external electric field E (when it is present). From electrostatic theory, the force on a charged particle immersed in an electric field E is

$$\mathbf{F}_e = q_p \mathbf{E} \tag{5.4}$$

where q_p is the charge on the particle. In the absence of an external component, the electric field affecting the interacting bodies results from the charges on their surfaces. In general, this is the sum of the forces due to the point charge interaction and the image force. However, Wang (1983b) showed that in the normal range of aerosol electric charges in the atmosphere, the image force could be neglected. Therefore, we consider only the point charge interaction in the following discussions. We shall also assume that the electric effect due to the electrically charged aerosol particle is negligible on the motion of the ice crystal. The electric field around the electrically charged ice crystal satisfies the condition

$$\mathbf{E} = -\nabla \Phi_e \tag{5.5}$$

where Φ_e is the electric potential around the crystal. Assuming that the space charge effect due to aerosol particles is again negligible, the electric potential satisfies the Laplace equation

$$\nabla^2 \Phi_e = 0 \tag{5.6}$$

The explicit form of the electric fields and potentials depend on the specific coordinate system and ice crystal habit. We will use ice plates as an example to demonstrate the effect of electric forces on the scavenging of aerosol particles by ice crystals.

Solving Eq. (5.1) yields the particle trajectory. The collision efficiency is again determined by

$$E = \frac{k}{k^*} \tag{5.7}$$

as discussed in Section 3 for the collisional growth of ice particles. We shall also assume that once collision occurs, the aerosol particle will stick with the ice crystal (i.e., retention efficiency is 1). With this assumption, the collision efficiency is the same as the collection efficiency (= collision efficiency × retention efficiency).

5.4.2. The Flux Model

For particles of $r < 0.5$ μm the effect of Brownian diffusion becomes important and the trajectory model is unsuitable for determining the scavenging efficiency. Instead, the flux model is used for this purpose. The following treatment follows mainly Wang (1985).

In this model we consider the scavenging of aerosol particles by snow crystals falling in the air due to the simultaneous action of Brownian motion of the aerosol particle (aerosol particle), electric and phoretic forces, inertial impaction, and turbulence. For particles of radii <0.5 μm the inertia term is usually very small and can be neglected. The turbulence effect also turns out to be insignificant (S. Grover, 1985; private communication) for particles of these sizes. Therefore, in the present model, only Brownian diffusion, phoretic forces, and electric forces will be considered. Similar considerations were taken by Wang et al. (1978) and Wang (1983b) for aerosol particle scavenging by small water drops, and the theoretical predictions were verified by the experiments of Wang and Pruppacher (1977), Leong et al. (1982), and Deshler (1982).

Under such assumptions the flux density of aerosol particle toward a stationary snow crystal of arbitrary shape is

$$\mathbf{j}_p = nB(\mathbf{F}_e + \mathbf{F}_{Th} + \mathbf{F}_{Df}) - D \nabla n \tag{5.8}$$

where n, B, and D are, respectively, the concentration, mobility, and diffusivity (assumed to be constant) of aerosol particle. $\mathbf{F}_e$, $\mathbf{F}_{Th}$, and $\mathbf{F}_{Df}$ represent the electrical, thermophoretic, and diffusiophoretic forces. Assuming steady-state conditions, the resulting continuity equation of aerosol particle concentration $\nabla \cdot \mathbf{j}_p = 0$ (which describes the conservation of aerosol particle in the space exterior to the snow crystal) leads to

$$B(\mathbf{F}_e + \mathbf{F}_{Th} + \mathbf{F}_{Df}) \cdot \nabla n - D \nabla^2 n = 0 \tag{5.9}$$

which is the well-known convective diffusion equation. Here we have used the fact that the forces involved are conservative and nondivergent. The latter condition does not hold for F_{Th} and F_{Df} when the snow crystal is not stationary but falling in the air, for then the temperature and vapor density fields are coupled with the air flow. A first-order correction to include the air flow effect may be made by introducing the mean ventilation coefficients $\bar{f}_h$, $\bar{f}_v$, and $\bar{f}_p$, which represent the enhancement factors due to air flow for the respective fluxes of heat, water vapor, and particle diffusion. The phoretic forces are then represented by $\bar{f}_h F_{Th}$ and $\bar{f}_v F_{Df}$. The electric force F_e is unaffected by the flow. The ventilation factors can be calculated using the following empirical formulas developed by Hall and Pruppacher (1976):

$$\bar{f}_v = 1 + 0.14X^2, \qquad X < 1.0 \tag{5.10}$$

$$\bar{f}_v = 0.96 + 0.28X, \qquad X \geq 1.0 \tag{5.11}$$

where X is defined as

$$X = N_{Sc,v}^{1/3} N_{Re,L^*}^{1/2} \tag{5.12}$$

Here $N_{Sc,v}$ is the Schmidt number of water in air and N_{Re,L^*} is the Reynolds number of snow crystals of characteristic dimension L^* defined as

$$L^* = \Omega/P \tag{5.13}$$

where Ω is the total surface area of the snow crystal and P is the perimeter of its area projected normal to the flow direction. It turns out that $\bar{f}_h \approx \bar{f}_v$ for typical atmospheric conditions and that is assumed in the present study. The above scheme has been used with success by Hall and Pruppacher (1976) to study heat and mass transfer of ice crystals in cirrus clouds. Newer ventilation coefficients obtained by Ji and Wang (1998), as introduced in Section 4, have not been applied to the scavenging problem yet.

Using such an approximation, we have

$$\nabla \cdot F = \nabla \cdot (F_e + \bar{f}_h F_{Th} + \bar{f}_v F_{Df}) = 0 \tag{5.14}$$

where F represents the sum of all three forces. Since the forces are conservative, we can express them as $F = -\nabla\phi$, where ϕ is a scalar potential, and from (5.14)

$$\nabla^2 \phi = 0 \tag{5.15}$$

The approximated convective-diffusion equation for particles surrounding a falling snow crystal is therefore

$$\bar{f}_p D \nabla^2 n - BF \cdot \nabla n = 0 \tag{5.16}$$

where $\bar{f}_p$ is the ventilation factor for the Brownian flux due to the motion of the snow crystal, which can be calculated using Eqs. (5.10)–(5.12) except that the Schmidt number in (5.12) now refers to that of the aerosol particle.

The boundary conditions necessary to solve Eq. (5.16) completely are

$$n = 0 \quad \text{at the snow surface}$$
$$n = n_\infty \quad \text{as } R \to \infty \tag{5.17}$$

which can be written as

$$n = 0 \quad \text{at } \phi = \phi_0$$
$$n = n_\infty \quad \text{at } R \to \infty \tag{5.18}$$

where ϕ_0 is the total force potential at the surface of the snow crystal and R is the distance from the surface. This set of boundary conditions simply says that once a particle hits the surface of the snow, it is retained there (i.e., the snow surface is a perfect sink) and that far away from the snow surface the concentration of aerosol particles remains a constant. A solution of (5.16) that satisfies the boundary conditions (5.18) is

$$n = n_\infty \left\{ \frac{\exp[B(\varphi_0 - \varphi)/D\bar{f}_p] - 1}{\exp[B\varphi_0/D\bar{f}_p] - 1} \right\} \tag{5.19}$$

The collection kernel is determined by integrating the flux density over the snow crystal surface:

$$K = -\frac{1}{n_\infty} \frac{\partial N}{\partial t} = -\left(\frac{1}{n_\infty}\right) \oint_s (n B \mathbf{F} - D\bar{f}_p \nabla n) \cdot d\mathbf{s} \tag{5.20}$$

where N is the total number of aerosol particles. Putting the solution (5.19) into (5.20), we obtain

$$K = -(1/n_\infty)/[\exp(B\varphi_0/D\bar{f}_p) - 1] \left[\oint_s n_\infty B \mathbf{F}\{\exp[B(\varphi_0 - \varphi)/D\bar{f}_p] - 1\} \cdot d\mathbf{s} \right.$$
$$\left. - \oint_s D\bar{f}_p(-n_\infty B/D\bar{f}_p) \exp[B(\varphi_0 - \varphi)/D\bar{f}_p] \nabla\varphi \cdot d\mathbf{s} \right] \tag{5.21}$$

Both integrals in (5.21) are to be evaluated on the surface of the snow crystal. Since $\varphi = \varphi_0$ at the surface, we immediately have

$$\exp[B(\varphi_0 - \varphi)/D\bar{f}_p] = \exp[0] = 1$$

and therefore the first integral in Eq. (5.21) vanishes. Consequently,

$$K = -(1/n_\infty)/[\exp(B\varphi_0/D\bar{f}_p) - 1] \left[\oint_s (-n_\infty B) \nabla\varphi \cdot d\mathbf{s} \right]$$
$$= \{B/[\exp(B\varphi_0/D\bar{f}_p) - 1]\} \oint_s \mathbf{F} \cdot d\mathbf{s} \tag{5.22}$$

We could have seen the vanishing of the first term in the integral of (5.20) sooner by noting that $n = 0$ at the surface, also implying that once the particles reach the surface, they stick there. In case particles rebound off the surface, this condition may not be quite accurate. Although there is no clear evidence of rebounding at present, this may happen at very low temperatures when the ice surface lacks a pseudo-liquid layer.

There are some questions regarding whether the model is independent of snow crystal shape. To clarify this point, we first note that throughout the whole derivation, including the verification given here, we rely solely on vector operations, which are independent of any particular coordinate system. This says that the results are applicable to any geometry considered. Second, the boundary conditions (5.18) are also written in a form independent of the coordinate system, and (5.19) satisfies these conditions without reference to the crystal shape.

The above shape-independence statement does not mean that the *numerical* values of n and K will be the same for all shapes of snow crystals. These values will be different because of the differences in φ and φ_0. We are merely saying that the form of the solution is shape-independent. The value of K, for example, depends on the capacitance C, which is a function of shape.

The origin of this shape independence comes from the *decoupling* of hydrodynamic drag force and other forces. In writing down (5.9), we have assumed that the hydrodynamic effects have been taken care of by the ventilation factors. Thus, (5.9) is essentially the convective diffusion equation for a stationary crystal except that each scavenging mechanism is now enhanced by a ventilation factor. In other words, we have assumed that the flux of aerosol particles toward a falling snow crystal due to a specific mechanism is equivalent to that toward a stationary snow crystal due to the same mechanism, except multiplied by a ventilation factor. The theoretical justification of this assumption has not been worked out yet. But the numerical results agree excellently with those computed from the much more involved hydrodynamic model and experimental results (Wang *et al.*, 1978). It is therefore justified in the practical sense.

The collection efficiency E can be determined according to

$$E = K/K^* = K/(Av_\infty) \tag{5.23}$$

where A and v_∞ are the cross-sectional area of the snow crystal perpendicular to the fall direction and the thermal velocity of the snow, respectively, and K^* is the geometric kernel. We have assumed that the sizes and terminal velocities of aerosol particles are very small compared to those of the snow crystals. This definition of E is essentially the same as that defined in Section 4.

The integral in Eq. (5.22) is

$$\int \mathbf{F} \cdot d\mathbf{s} = \int \mathbf{F}_e \cdot d\mathbf{s} + \int \bar{f}_h \mathbf{F}_{Th} \cdot d\mathbf{s} + \int \bar{f}_v \mathbf{F}_{Df} \cdot d\mathbf{s} \tag{5.24}$$

The explicit expressions of the terms on the right-hand side can be determined by the so-called electrostatic analog theory (see, for example, p. 448 of Pruppacher and Klett, 1978) which utilizes the Gauss theorem. Using this theorem immediately leads to

$$\int \mathbf{F}_e \cdot d\mathbf{s} = 4\pi C q(V_s - V_\infty) = 4\pi Q q \tag{5.25}$$

where C is the capacitance of the snow crystal under consideration, V_s is its surface electric potential, V_∞ is the electric potential at infinity, and Q and q are the electrostatic charges of snow crystals and aerosol particles, respectively. We have ignored the image force here. We have used the fact that $Q = 4\pi C(V_s - V_\infty)$. Here we have assumed that particles behave like point charges and the crystal surface represents an equipotential surface. Similarly,

$$\int \mathbf{F}_{Th} \cdot d\mathbf{s} = 4\pi C \bar{f}_h Z_{Th}(T_s - T_\infty) \tag{5.26}$$

where

$$Z_{Th} = \frac{[12\pi \eta_a r(k_a + 2.5k_p N_{Kn})k_a]}{[5(1 + 3N_{Kn})(k_p + 2k_a + 5k_p N_{Kn})P]} \tag{5.27}$$

and

$$\int \mathbf{F}_{Df} \cdot d\mathbf{S} = 4\pi C \bar{f}_v Z_{Df}(\rho_{v,s} - \rho_{v,\infty}) \tag{5.28}$$

where

$$Z_{Df} = \frac{[6\pi \eta_a r(0.74 D_V M_a)]}{[(1 + \alpha N_{Kn})M_w \rho_a]} \tag{5.29}$$

In the above equations, N_{Kn} is the Knudsen number ($= \lambda_a/r$, where λ_a is the mean free path of air molecules) and α is the Stokes–Cunningham slip correction factor (see Pruppacher and Klett, 1997).

For very small aerosol particles (say, $r < 0.1$ μm), Eq. (5.29) needs modification. However, the overall effect of such modification is expected to be small since Brownian diffusion will be the predominant mechanism of collection for small aerosol particle.

Using Eqs. (5.25)–(5.27), the collection kernel (5.22) for aerosol particles scavenged by a snow crystal can be written as

$$
\begin{aligned}
K &= -4\pi B \frac{[Qq + C\bar{f}_h Z_{Th}(T_s - T_\infty) + C\bar{f}_v Z_{Df}(\rho_{v,s} - \rho_{v,\infty})]}{[\exp(B\phi_0/D\bar{f}_p) - 1]} \\
&= \frac{4\pi B\phi_0 C}{\{[\exp(B\phi_0/D\bar{f}_p) - 1]\}} \tag{5.30}
\end{aligned}
$$

since the terms in the numerator are exactly $\phi_0 C$. The collection efficiency E can be determined using the relation

$$E = \frac{4\pi B \phi_0 C}{\{[\exp(B\phi_0 / D\bar{f}_{\mathrm{p}}) - 1] A v_\infty\}} \tag{5.31}$$

The full form of the surface potential ϕ_0 is

$$\phi_0 = \left[\frac{Qq}{C}\right] + \frac{\bar{f}_{\mathrm{h}}[12\pi\eta_{\mathrm{a}} r (k_{\mathrm{a}} + 2.5 k_{\mathrm{p}} N_{\mathrm{Kn}}) k_{\mathrm{a}}]}{[5(1 + 3N_{\mathrm{Kn}})(k_{\mathrm{p}} + 2k_{\mathrm{a}} + 5k_{\mathrm{p}} N_{\mathrm{Kn}}) P]}$$
$$+ \frac{\bar{f}_v [6\pi\eta_{\mathrm{a}} r (0.74 D_{\mathrm{V}} M_{\mathrm{a}})]}{[(1 + \alpha N_{\mathrm{Kn}}) M_{\mathrm{w}} \rho_{\mathrm{a}}]} \tag{5.32}$$

Once the efficiencies from the two models are determined, they can be merged to give the efficiencies over the whole aerosol particle size spectrum. The merging is not done arbitrarily, but according to the best matching of the two curves. Figure 5.1 shows an example (Miller, 1988). In this example, the two curves virtually coincide for $0.3 < r < 0.7$ μm, and the merging is indeed very natural and smooth. The overlapping ranges are different for different colliding pairs and atmospheric conditions, but all of them merge smoothly.

The merger is almost guaranteed to be smooth for a good reason. As we will see in more detail later, the dominant scavenging mechanism for $r < 0.1$ μm is Brownian

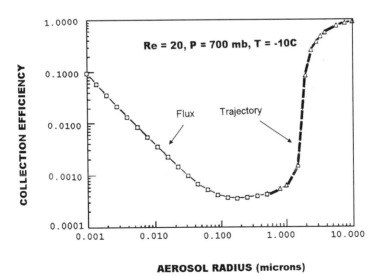

FIG. 5.1. Matching of the FLUX and TRAJECTORY models for Re = 20 with RH = 95% and no electrostatic force.

diffusion whereas that for $r > 1.0$ μm is inertial impaction. In the range of $0.1 < r < 1.0$ μm, Brownian diffusion and inertial impaction are negligible compared to the phoretic and electric forces in determining the collection efficiency. The representations of these forces are nearly the same in these two models (the only difference is in the ventilation effect, which is very close to the effect of the flow field). Hence the computed efficiencies are about the same.

5.5. Efficiencies of Ice Plates Collecting Aerosol Particles

Martin *et al.* (1980a,b) studied the scavenging of aerosol particles by ice plates. They approximated the hexagonal ice plates by thin oblate spheroids of axis ratio 0.05. The flow fields were those obtained by Pitter *et al.* (1973). Note that the relevant coordinate system for this problem is the oblate spheroidal coordinate system. Table 5.1 shows the dimensions and other characteristics of the ice plates.

The aerosol particle radii considered were $0.001 \leq r \leq 10$ μm. The ice crystal plates had radii (i.e., semimajor axes) of $a_c = 50.6$, 87.9, 112.8, 146.8, 213, 289, 404, and 639 μm, corresponding to Reynolds numbers $N_{Re} = 0.1$, 0.5, 1.0, 2.0, 5.0, 10.0, 20, and 50, respectively, at 700 mbar and $-10°$C. In addition to this pressure level, both models were evaluated for the levels 1000 mbar, $0°$C; 900 mbar, $-5°$C; and 600 mbar, $-20°$C. Owing to the particular choice of corresponding pressure and temperature, the Reynolds numbers corresponding to the above given crystal sizes were, with sufficient accuracy, the same at all pressure–temperature levels considered. At each pressure level we considered four relative humidities (RH)$_i$ (with respect to ice): namely, (RH)$_i = 100$, 95, 75, and 50%. The values chosen for $k_a(p, T)$, $\lambda_a(p, T)$, $D_{va}(p, T)$, and $\eta_a(T)$ were those recommended by Pruppacher and Klett (1978). The values for ρ_a were those given by the Smithsonian Meteorological Tables. The bulk densities of the aerosols were $\rho_p = 1.0$, 1.5, 1.75, 2.0, and 5 g cm^{-3}. In the present computations we also assumed that $f_h \approx f_v$, where f_v is the ventilation coefficient for mass transport. In evaluating f_p we assumed that its functional dependence on the Reynolds and Schmidt numbers N_{Re} and N_{Sc} is the same as that given by Hall and Pruppacher (1976) for f_v, except that now instead of $N_{Sc,v} = \nu_a/D_{v,a}$ we used $N_{Sc,p} = \nu_a/D_{p,a}$, where $D_{p,a}$ is the diffusivity of the aerosol particles in air. Values for $D_{p,a}$ and justifications for both of the above assumptions are given by Pruppacher and Klett (1978). The thermal conductivity of the aerosol particle material was assumed to be $k_p = 4.19 \times 10^{-1}$ J cm^{-1} s^{-1}°C^{-1}. For evaluating the phoretic forces, a uniform ice crystal temperature was assumed, considering the thinness of the ice crystals and the relatively high heat conductivity of ice.

Little is known about the surface charge Q_a on platelike ice crystals. However, the scant information available provided bounds from which it was determined that the surface charge on platelike ice crystals in strongly electrified clouds may

TABLE 5.1 REYNOLDS NUMBER OF OBLATE SPHEROID OF ICE
CORRESPONDING TO A GIVEN SEMIMAJOR AXIS (RADIUS) AS A FUNCTION OF
PRESSURE–TEMPERATURE LEVEL

Reynolds number of available flow field	a_c (μm)	p (mbar)	T (°C)	N_{Re}
0.1	50.6	600	−20	0.097
		700	−10	0.10
		900	−5	0.111
		1000	0	0.114
0.5	87.9	600	−20	0.486
		700	−10	0.499
		900	−5	0.553
		1000	0	0.568
1.0	112.8	600	−20	0.973
		700	−10	0.999
		900	−5	1.01
		1000	0	0.14
2.0	146.8	600	−20	1.95
		700	−10	1.998
		900	−5	2.21
		1000	0	2.27
5.0	213	600	−20	4.88
		700	−10	5.00
		900	−5	5.54
		1000	0	5.7
10	289	600	−20	9.73
		700	−10	9.99
		900	−5	11.05
		1000	0	11.36
20	404	600	−20	19.32
		700	−10	20.02
		900	−5	22.15
		1000	0	22.8
50	639	600	−20	48.75
		700	−10	50.00
		900	−5	55.3
		1000	0	56.9

be represented by

$$|Q_a| = |q_a|a_c^2 = 2a_c^2 \tag{5.33}$$

An analogous law was shown Wang *et al.* (1978) to hold for spherical particles. Thus we assumed for strongly electrified clouds $|Q_r| = |q_r|r^2 = 2r^2$.

It also appears from the studies cited above that platelike crystals are predominantly negatively charged. In order to test the effect of electric charges on the scavenging of aerosol particles by ice crystals, we considered both strongly and weakly electrified clouds, and therefore investigated the charge effect for $|q_a| = |q_r| = 0.14, 0.20, 0.40, 1.0, 1.4$, and 2.0 esu cm^{-2}, assuming that the crystals were negatively charged and the aerosol particles were positively charged:

$$q_a = \frac{Q_a}{a_c^2} \quad \text{and} \quad q_r = \frac{Q_r}{r^2}$$

Obviously, the smallest charge an aerosol particle can carry is $Q_r = 4.8 \times 10^{-10}$ esu, which is equal to one electron charge. Smaller particles carry no charge. Thus, it appears that our formulation $Q_r = q_r r^2$ applies only to aerosol particles of $r \geq (4.8 \times 10^{-10}/q_r)^{1/2}$, i.e., to aerosol particles of $r \geq 0.2\ \mu$m, if we assume $q_r = 2.0$ esu cm^{-2}. However, since the trajectory model considered only particles of $r > 0.1\ \mu$m while the flux models considered particles of $0.001 \leq r \leq 0.1\ \mu$m, the above restrictions apply only to the latter. On the other hand, the flux model did not consider the motion of individual particles but rather the flux of a *whole assembly of particles,* some of which carry zero charge while others carry $1, 2, \ldots$ electron charges. Therefore, we assumed that, in the mean, the aggregate electric charge for the particles affecting the scavenging of the particle population could be given by $Q_r = q_r r^2$.

Examples of the computed results are given in Figures 5.2 to 5.5. These figures give the efficiency with which electrically charged and uncharged aerosol particles

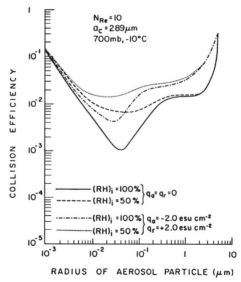

FIG. 5.2. Efficiency with which aerosol particles collide with a simple planar ice crystal of radius $a_c = 289\ \mu$m and Reynolds number Re $= 10$, in air of 700 mbar, $-10°$C, and of relative humidity (RH)$_i$ (with respect to ice) of 50, 75, 95, and 100%; for $\rho_p = 2$ g cm^{-3}, and for $q_a = q_r = 0$ and $q_a = q_r = 2.0$ esu cm^{-2}, where $q_a = Q_a/a_c^2$ and $q_r = Q_r/r^2$.

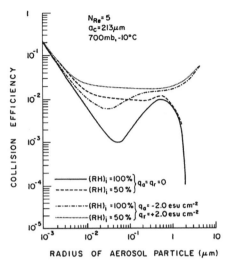

FIG. 5.3. Efficiency with which aerosol particles collide with a simple planar ice crystal of radius $a_c = 213$ μm and Reynolds number Re = 5, in air of 700 mbar, $-10°$C, and of relative humidity (RH)$_i$ (with respect to ice) of 50, 75, 95, and 100%; for $\rho_p = 2$ g cm^{-3}, and for $q_a = q_r = 0$ and $q_a = q_r = 2.0$ esu cm^{-2}, where $q_a = Q_a/a_c^2$ and $q_r = Q_r/r^2$.

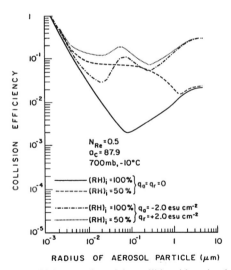

FIG. 5.4. Efficiency with which aerosol particles collide with a simple planar ice crystal of radius $a_c = 87.9$ μm and Reynolds number Re = 0.5, in air of 700 mbar, $-10°$C, and of relative humidity (RH)$_i$ (with respect to ice) of 50, 75, 95, and 100%; for $\rho_p = 2$ g cm^{-3}, and for $q_a = q_r = 0$ and $q_a = q_r = 2.0$ esu cm^{-2}, where $q_a = Q_a/a_c^2$ and $q_r = Q_r/r^2$.

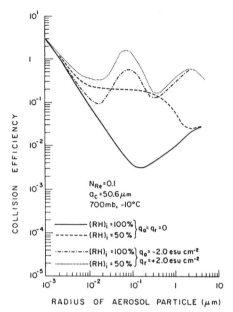

FIG. 5.5. Efficiency with which aerosol particles collide with a simple planar ice crystal of radius $a_c = 50.6\,\mu$m and Reynolds number Re $= 0.1$, in air of 700 mbar, $-10°$C, and of relative humidity $(RH)_i$ (with respect to ice) of 50, 75, 95, and 100%; for $\rho_p = 2$ g cm^{-3}, and for $q_a = q_r = 0$ and $q_a = q_r = 2.0$ esu cm^{-2}, where $q_a = Q_a/a_c^2$ and $q_r = Q_r/r^2$.

of $0.001 \leq r \leq 10\,\mu$m are captured by electrically charged and uncharged ice crystal plates of various radii in air of various humidities at $-10°$C and 700 mbar. The results at the other pressure–temperature levels under consideration differed only insignificantly from those at $-10°$C and 700 mbar. We attributed this finding to our particular choice of pressure–temperature level, which in combination affected η_a, D_{va}, ρ_a, and k_a in such a manner that the pressure- and temperature-sensitive contributions to the phoretic and hydrodynamic forces compensated each other. Other combinations of pressure and temperature may well change the present curves.

The most significant feature of the curves in Figures 5.2 to 5.5 is the predominant minimum in the collision efficiency E for aerosol particles of radius between $r = 0.01\,\mu$m and $r = 0.1\,\mu$m. Analogous to the particle scavenging behavior of water drops (Wang *et al.*, 1978), this result can be explained on the basis of Brownian diffusion, which is increasingly responsible for particle scavenging as the particle radius increases above 0.1 μm. However, it is worth noting that the minimum (termed the Greenfield gap by Wang *et al.*, 1978) for particle scavenging by ice crystal plates appears at aerosol particle radii one order of magnitude smaller than those at which the minimum appears for water drops. This result is caused by the unusual properties of the ice crystal rim as a trap for the aerosol particles. In contrast to the air flow past a spherical drop, air flow past a thin falling ice plate

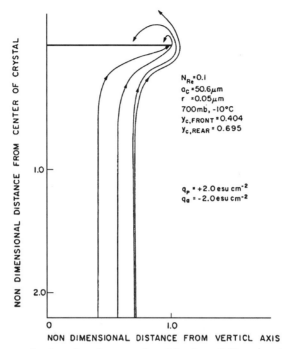

FIG. 5.6. Computed trajectories of an aerosol particle of $r = 0.05$ μm moving around an ice plate of $a_c = 50.6$ μm (Re = 0.1), for $q_a = q_r = 2.0$ esu cm^{-2}. Note that the capture of aerosol particles takes place at the ice crystal rim, and also may take place on the rear side of the ice crystal.

exhibits strong horizontal flow components on its lower side, recurving sharply to become more or less vertical near the crystal's edge, with the streamlines strongly crowding near the crystal tip as explained by Pitter *et al.* (1973). This flow behavior causes aerosol particle trajectories of the type described in Figure 5.6. These demonstrate that the ice crystal rim is a preferred capture site for aerosol particles. Thus, the capture of aerosol particles is limited by the air flow past the scavenging body to much smaller aerosol particle sizes for thin ice plates than for droplets. For aerosol particles captured by such plates, we observed no annular behavior of the type found by Pitter and Pruppacher (1974) for drops captured by ice plates.

It is evident from Figures 5.2 to 5.5 that, as in the case of particle scavenging by drops, particle scavenging by ice crystal plates is most strongly affected by phoretic forces in the Greenfield gap. The phoretic effects are quite small for aerosol particles if $r > 1$ μm or $r < 0.01$ μm, but are important if $0.01 \leq r \leq 1$ μm. Note also from these figures that the phoretic effects become stronger for smaller ice crystals. Obviously, the smaller the ice crystal, the smaller its Reynolds number and therefore the smaller the particle-deflecting effect of the flow field beneath the crystal.

Like the phoretic effects, the electrical effects on particle scavenging are negligible for $r < 0.01\ \mu m$. However, they are very pronounced for $0.01 \le r \le 10\ \mu m$, depending on the size of the ice crystal. Thus, the collision efficiency for crystals of $a_c = 639$, 404, and 289 μm is raised by as much as an order of magnitude in the range of $0.01 \le r \le 5\ \mu m$ if $|q_a| = |q_r| = 2.0$ esu cm^{-2}.

We also note from Figures 5.2 to 5.5 that the phoretic effects on scavenging are less noticeable if the aerosol particles and ice crystals are electrically charged. These figures show further that the smaller the Reynolds number of the ice crystal, i.e., the smaller the particle-deflecting hydrodynamic forces beneath the crystal, the more the collision efficiency is enhanced by the electric charges present.

Some particularly strong electric effects are noted for particles of $r > 1\ \mu m$ and ice crystals of $a_c = 213$, 146.8, and 112.8 μm. If crystals of these sizes are *electrically uncharged*, their collision efficiency rapidly decreases to zero as r becomes larger than 1 μm. In fact, no particles are collected if $r > 2\ \mu m$. Trajectory analysis shows that the reason for this behavior is that, at these relatively low Reynolds numbers, the approach velocity of the aerosol particle to the ice crystal is sufficiently small to afford the strong horizontal hydrodynamic deflecting forces beneath the ice crystal sufficient time to move any aerosol particle of $r > 2\ \mu m$ around the crystal, creating the collision efficiency of zero. However, if the ice crystal and aerosol particles are *electrically charged*, with $|q_a| = |q_r| = 2.0$ esu cm^{-2}, the collision efficiency becomes finite and in fact quite large, being raised to a value above 10^{-2} by the electric charges.

A further dramatic change in the collision behavior of ice crystal plates is noted if $N_{Re} < 1$ (see Figs. 5.4 and 5.5). We note that at these very low Reynolds numbers, aerosol particles of $r > 1\ \mu m$ are again captured. Analysis of the velocity field around the falling crystal shows that this behavior stems from a pronounced decrease of the horizontal, particle-deflecting velocity. In fact, the deflecting force for $N_{Re} < 1$ becomes so low that, despite the small approach velocity, the particle cannot escape a collision with the crystal. Nevertheless, the increase in E due to the presence of electric charges is considerable over the whole particle size range. In fact, with decreasing r, the collision efficiency decreases unexpectedly to a local minimum near $r \approx 0.5\ \mu m$ and subsequently increases with further decrease in r to a local maximum near $r \approx 0.05\ \mu m$ as r decreases further. Trajectory analysis illustrated in Figure 5.6 shows that this effect is due to capture of the charged aerosol particles on the *rear side* of the charged ice crystal. Figures 5.4 and 5.5 show that with even further decrease in particle size, the collision efficiency decreases again as the electric forces are now rapidly decreasing in comparison to the hydrodynamic forces, tending to move the particle around the crystal. The final increase in collision efficiency for $r < 0.01\ \mu m$ is due to the effects of Brownian diffusion. In fact, the Brownian motion is so efficient for such small particles that $E > 1$ for very small particles. This is a consequence of the effective collision kernel K becomes greater than the geometric kernel K^* [see Eq. (5.7)]; i.e., even

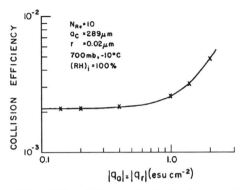

FIG. 5.7. Variation of the collision efficiency with electric charge on the crystal and aerosol particle for $a_c = 289$ μm and $r = 0.02$ μm.

those particles lying outside of the volume swept out by the ice crystal come to collide with the crystal.

In Figures 5.7 and 5.8 the variation of the collision efficiency is plotted as a function of the charge $|q_a| = |q_r|$ on the ice crystal and aerosol particle, respectively, where

$$q_a = \frac{Q_a}{a_c^2} \quad \text{and} \quad q_r = \frac{Q_r}{r^2}$$

We note that for both aerosol particle sizes considered and for an ice crystal of radius $a_c = 289$ μm, electric charges begin to noticeably affect the capture of aerosol particles if $|q_a| = |q_r| > 0.4$. This charge is considerably below the mean charge on particles in thunderstorm clouds. Figure 5.9 shows that for an ice crystal of $a_c = 404$ μm the collision efficiency is significantly affected even though the charge is as low as $|q_a| = |q_r| = 2.0$ esu cm^{-2}.

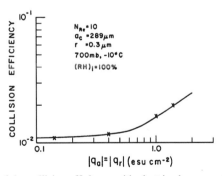

FIG. 5.8. Variation of the collision efficiency with electric charge on the crystal and aerosol particle for $a_c = 289$ μm and $r = 0.3$ μm.

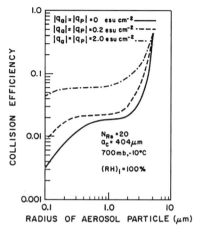

FIG. 5.9. Variation of the collision efficiency with aerosol particle radius for various electric charges on the ice crystal and aerosol particle for $a_c = 404$ μm.

5.6. Efficiencies of Columnar Ice Crystals Collecting Aerosol Particles

Using the methods described above, Wang and Pruppacher (1980a) and Miller and Wang (1989) determined the efficiencies with which aerosol particles are collected by columnar ice crystals. The discussions below closely follow Miller and Wang (1989). In this study, the air flow past ice columns is approximated by that past infinitely long cylinders. This approximation is valid for larger ice crystals, but may underestimate the efficiencies for smaller ice crystals especially for Re < 1 (see Sec. 4). Fortunately, the underestimation is not severe and does not affect the conclusions drawn below.

Solutions to Eqs. (5.1) and (5.31) were computed for seven columnar ice crystals with Reynolds numbers, radii, lengths, densities, and terminal velocities as listed in Table 5.2. The results presented here correspond to an atmospheric pressure of 600 mbar and an ambient temperature of $-20°C$ unless otherwise

TABLE 5.2 COLUMNAR ICE CRYSTAL CHARACTERISTICS

N_{Re}	r_c (μm)	l_c (μm)	ρ_c (g/cm^3)	V_∞ (cm/s)	$Q_{c/1}$(esu/cm)
0.5	32.7	93.3	0.6	12.21	-2.139×10^{-5}
0.7	36.6	112.6	0.6	15.26	-2.679×10^{-5}
1.0	41.5	138.3	0.6	19.22	-3.445×10^{-5}
2.0	53.4	237.4	0.6	29.87	-5.703×10^{-5}
5.0	77.2	514.9	0.6	51.65	-1.192×10^{-5}
10.0	106.7	1067.1	0.6	74.76	-2.277×10^{-5}
20.0	146.4	2440.0	0.6	108.98	-4.287×10^{-5}

noted. Values for $\rho_a(P, T)$ are from the Smithsonian Meteorological Tables. Formulas for other key variables are $k_a(T) = (5.69 + 0.017T°C) \times 10^{-5}$, $k_p(T) = (3.78 + 0.020T°C) \times 10^{-5}$, $\lambda(P, T) = 6.6 \times 10^{-6}$ cm $(1013.15/P)(T°K/293.15)$, $D_{va}(P, T) = 0.211(T°K/273.15)^{1.94}$ $(1013.25/P)$, and $\eta_a(T) = (1.718 + 0.0049T - 1.2 \times 10^{-5}T^2) \times 10^{-4}$, $T(°C) < 0$, each from Pruppacher and Klett (1978). The unit for the pressure P is the millibar. The bulk densities of the ice crystals and aerosol particles are 0.6 and 2.0 g cm^{-3}, respectively, unless otherwise noted.

Figure 5.10 indicates collection efficiencies for particles $0.001 \leq r_p \leq 10.0\ \mu$m by columnar ice crystals with $0.5 \leq N_{Re} \leq 20.0$. The relative humidity is 95% with respect to ice (all relative humidities presented here are with respect to ice). It is seen that particles less than $0.9\ \mu$m in radius have increasing collision efficiencies for decreasing N_{Re}. Particles with $r_p > 2.0\ \mu$m exhibit the opposite behavior. The increase in efficiency for decreasing N_{Re} for $r_p > 2.0\ \mu$m is due to the increased contribution from the inertia term. In this case, as N_{Re} increases, the corresponding flow speed u increases, as does $(u - v_d)$ in F_d.

Figures 5.11a–c show examples of the collection efficiencies for $N_{Re} = 0.5$, 1.0, and 20.0 for aerosol particles of $0.001 \leq r_p \leq 10.0\ \mu$m with different relative humidities as plotted in curves 1 (95%), 2 (75%) and 3 (50%), both without (solid curves, no suffix) and with charge (dotted curves, suffix "e"). In these figures, gravitational, inertial, electrostatic, thermophoretic, and diffusiophoretic forces as

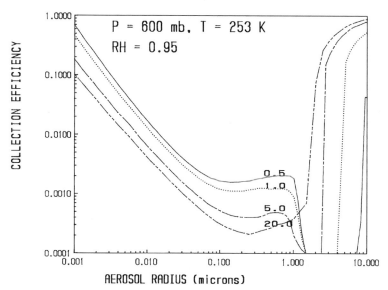

FIG. 5.10. Variation of the collection efficiency for aerosol particles by columnar ice crystals of various Reynolds numbers at 600 mbar and $-20°C$.

a

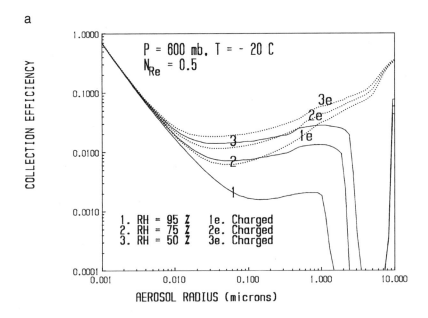

b

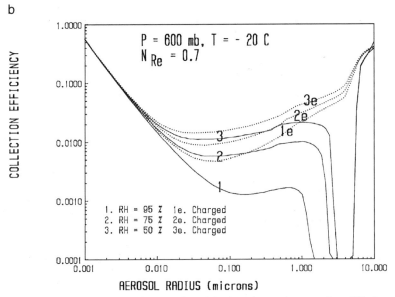

FIG. 5.11. Collection efficiency for aerosol particles by columnar ice crystals at 600 mbar and −20°C versus aerosol particle radius. Labels 1, 2 and 3 represent RH = 95%, 75%, and 50%, respectively. The suffix "e" denotes electrostatic charge of $q = 2.0$ esu cm^{-2}. The Reynolds numbers of the ice crystal are (a) 0.5, (b) 0.7, (c) 1.0.

C

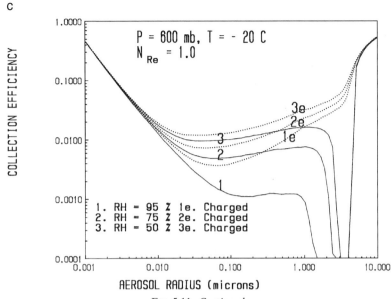

FIG. 5.11. *Continued.*

well as Brownian diffusion are present. Our choice of the three humidity values highlights the temperature and water vapor dependencies of thermo- and diffusiophoretic forcing mechanisms in these figures. It is seen that phoretic forcing occurs in subsaturated air for particles of $0.01 \leq r_p \leq 2.5$ μm. The magnitude of this forcing increases with decreasing relative humidity and for decreasing N_{Re}. It should be noted that thermophoresis is toward the ice crystal surface when it is cooler than the ambient surroundings, while diffusiophoresis acts outward from the crystal surface under the same condition (for a falling columnar ice crystal). When $r_p < 1.0$ μm, $F_{Th} > F_{Df}$, and vice versa for $r_p > 1.0$ μm. The greatest net inward phoretic forcing ($F_{Th} + F_{Df}$) is seen to occur for particles slightly less than 1.0 μm.

The effect of electric charges is shown by curves labeled with the suffix "e" after the curve number. As in the case of ice plates, in each instance $q = 2$ esu cm^{-2}, corresponding to measured electrostatic charge on drops in a thunderstorm (Takahashi, 1973). The presence of electrostatic charges is seen to increase the collection efficiency, markedly for high relative humidities and less so for low relative humidities, for which the efficiency is already much higher than in the moist scenarios. The electrostatic effect increases in proportion to the squares of increasing aerosol particle and ice crystal radii.

The observed zero scavenging zone (ZSZ) within the Greenfield gap in Figure 5.11 occurs as the sum of the radially directed forces $F_g + F_d + F_{Th} + F_{Df}$ approaches zero. Nonzero scavenging values appear as ($F_{Th} + F_{Df}$) is increased in

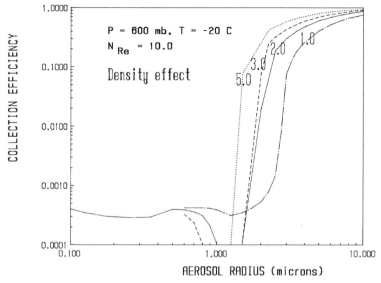

FIG. 5.12. Collection efficiency as a function of aerosol particle density at 1.0, 2.0, 3.0, and 5.0 g cm^{-3} at 600 mbar, $-20°$C, and 95% RH.

magnitude. As N_{Re} decreases, the ZSZ widens in the direction of increasing particle size; at 600 mbar with 95% relative humidity, the ZSZ covers radii between 1 and 1.5 μm for $N_{Re} = 10$, but between 1.5 and 2.5 μm for $N_{Re} = 5$ and between 1.5 and 7.5 μm for $N_{Re} = 0.5$.

Figure 5.12 illustrates the dependence of the collision efficiency on particle density. Particle density effect is not important for $r_p < 0.5$ μm flux calculations; however, it becomes very significant for $r_p > 0.5$ μm trajectory calculations. In general, increasing particle density increases collision efficiency for $r_p \geq 1.5$ μm. However, the curves are crossing each other for $r_p < 1.5$ μm, and the efficiency decreases with increasing density, presumably because of the weakening inertial effects.

Figure 5.13 shows the effect of temperature on scavenging efficiency. Increasing temperature increases the efficiency for $r_p < 1.0$ μm, but the opposite is true for $r_p > 2.0$ μm. Increasing temperature for minimal phoretic forcing will bridge the ZSZ. Variations in pressure are in the same direction as for temperature, as illustrated in Figure 5.14, showing the collision efficiency for a columnar ice crystal with $N_{Re} = 10$ for several pressure–temperature levels in the atmosphere. Figure 5.15 compares the columnar ice crystal aerosol scavenging at 700 mbar with the drop aerosol scavenging at 900 mbar. The ice crystal–aerosol results are for a temperature of 263K while the drop–aerosol results are for 283K; both are at 95% relative humidity, but taken over ice and liquid water, respectively. These results, which are plotted versus their equivalent geometric kernels K^*, indicate that columnar ice crystals scavenge more efficiently than drops across the Greenfield

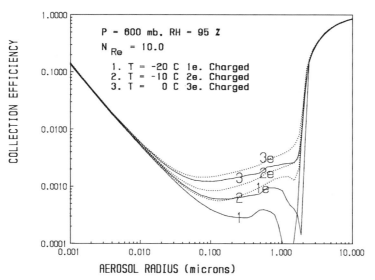

FIG. 5.13. Collection efficiency of aerosol particles by a columnar ice crystal with Re = 10.0 at 600 mbar, −20°C, and 95% RH. Curves represent temperature and electrostatic charge: (1) −20°C, (2) −10°C, (3) 0°C. The suffix "e" denotes electrostatic charge of $q = 2.0$ esu cm^{-2}.

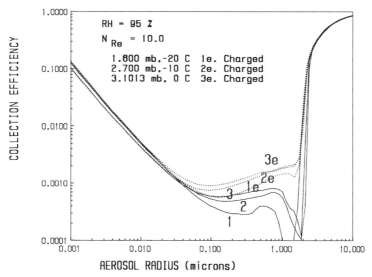

FIG. 5.14. Collection efficiency as a function of pressure–temperature levels by a columnar ice crystal with Re = 10.0 and 95% RH. Curves represent temperature and electrostatic charge: (1) 600 mbar, −20°C; (2) 700 mbar, −10°C; (3) 1013 mbar, 0°C. The suffix "e" denotes electrostatic charge of $q = 2.0$ esu cm^{-2}.

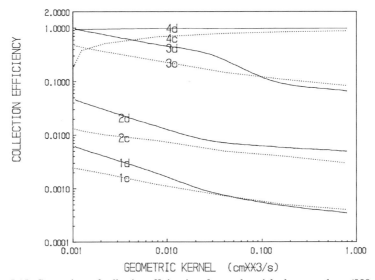

FIG. 5.15. Comparison of collection efficiencies of aerosol particles by water drops (900 mbar, 10°C, solid) and columnar ice crystals (700 mbar, -10°C, dotted) for particle radius (1) 0.1 μm, (2) 0.01 μm, (3) 0.001 μm, and (4) 10 μm. Labels c and d represent crystal and drop, respectively. RH is 95%.

gap, except in the region of the ZSZ. This zone, which is also seen in previous results for drop aerosol scavenging (Wang *et al.*, 1978) as well as ice plate aerosol scavenging (Martin *et al.*, 1980a,b), is wider and occurs over a greater geometric range for columnar ice crystals.

In the above, we have presented a combined numerical and analytical model that investigates the scavenging efficiencies of aerosol particles by columnar ice crystals. We have shown that relative humidity, temperature, pressure, and electrostatic charge variations can alter the collection efficiency across the Greenfield gap. The foregoing study quantitatively indicates increasing efficiency for decreasing relative humidity and/or increasing temperature, pressure, and electrostatic charge. The presence of a zero scavenging zone for flows past ice crystals with Re < 20 was observed for particles 1.5 to 2.0 μm, and this is seen to be a cancellation of radially directed forces. This gap widens with increasing pressure and with decreasing temperature.

5.7. Comparison of Collection Efficiency of Aerosol Particles by Individual Water Droplets, Ice Plates, and Ice Columns

5.7.1. Introduction

Clouds and precipitation may consist of water drops, ice crystals, or both. All three cases exist in any season. Clouds in wintertime may consist of liquid drops and clouds in summertime may contain ice particles. Thus a complete understanding of precipitation scavenging, as defined in the beginning of Section 5.2, would necessarily involve the study of both rain and ice scavenging. The main question of interest here is how the relative scavenging efficiencies of water drops and ice crystals compare. Comments in past studies suggest that snow is a more efficient particle scavenger than rain (Carnuth, 1967; Magono et al., 1974; Murakami et al., 1985a,b). The comparisons were based on the equivalent liquid water contents. For data derived from field observations, this is probably the only way to compare the efficiencies. Recent theoretical studies and laboratory experiments (Slinn and Hale, 1971; Wang and Pruppacher, 1977; Wang et al., 1987; Martin et al., 1980a,b; Murakami et al., 1985a,b; Sauter and Wang, 1989; Miller and Wang, 1989) permit comparison between efficiencies by individual collectors.

5.7.2. The Basis for Comparison

One question has often been asked when comparing the scavenging efficiencies of rain and snow: What is the basis for comparison? This is a valid question, one that should be answered before we proceed. The most straightforward method of comparison is based on the size of the collectors, this method is easily understood when the collectors have the same shape (e.g., both are spheres or hexagonal plates) but leads to confusion when collectors are of different shapes. For example, a raindrop whose diameter is the same as that of a hexagonal ice plate will normally fall much faster than the plate. If the drop collects more aerosol particles, it may simply be that the drop has traveled a longer distance and hence has had more chance to collect particles. This is also true for ice particles of the same size but different shapes. Another method of comparison, already mentioned in the previous section, is based on the equivalent liquid water contents or equivalent rainfall rates (see, for example, comments in Wang and Ji (2000)). This may be useful for field observational data, but in view of the difficulty in separating various kinds of collector particles and mechanisms, it is to be noted that this method considers the integrated effect of the whole collector size (and often also shape) distribution. The real ability of each collector particle is blurred by that of other particles. Thus, in the present study only the collection efficiencies of individual collectors will be compared. Such a comparison will be especially useful in elucidating the physical mechanisms of scavenging and their relative contributions in the collection and removal of atmospheric aerosol particles. The collection efficiencies are measures of the "ability" of the collectors to collect particles; thus we should use a fair basis for comparison. We propose here to use K^*, the geometric kernel introduced in Section 4. There have also been suggestions that the term "geometric sweepout volume per unit time" should be used, to avoid confusion with the term "scavenging kernel." The relation between K^* and the conventional scavenging kernel K (Pruppacher and Klett, 1978) can be made clear from the following expression:

$$E = K/K^*$$

i.e., the scavenging kernel is simply the collection efficiency times K^*. Alternatively, the collection efficiency is simply the ratio of K to K^*. Thus, for example, K^* for a drop of radius a is

$$K^* = \pi a^2 V_\infty$$

while K^* for a columnar ice crystal of length L and radius a is

$$K^* = 2aL V_\infty$$

where V_∞ represents the terminal fall velocity of the collector. In this definition of K^*, we have assumed that all hydrometeors fall with their largest dimensions oriented horizontally. This assumption is valid for relatively small hydrometeors, which are being considered in the present study. For larger hydrometeors, the orientation may change with time, whereupon the definition of K^* must be modified, but this will not be considered here. The rationale of using K^* as the basis for comparing collection efficiencies is as follows. Since we are comparing collectors of different shapes, we should compare their efficiencies under the condition that they are given the same chance to be exposed to the same amount of aerosol particles per unit time. We shall assume that the aerosol concentration is uniform, which is necessary for a fair comparison. Under this condition, the same volume of air would contain the same amount of aerosol particles. Thus, when we compare the collection efficiencies for collectors with the same K^*, we are exposing the collectors (even though of different shapes) to the same amount of aerosol particles per unit time. The one that collects more particles does so because it is really more efficient in capturing particles. Hence using K^* as the basis for comparison is "fair." Apart from Wang and Pruppacher (1980a), Podzimek (1987) also used the same technique for comparing collection efficiencies.

5.7.3. Data Sources of Collection Efficiencies

The "data" used for the present comparison are results derived from calculations based on the models developed by Wang et al. (1978) for drops, as well as the results presented in the previous two sections for ice particles. Ideally, experimental data sets such as those obtained by Wang and Pruppacher (1977), Lai et al. (1978), Prodi (1976), Murakami et al. (1985a,b), Sauter and Wang (1989), and Song and Lamb (1992) should be the ones used for the comparison. However, at present the experimental data are not complete enough to form continuous data sets, and the available sets do not have substantially overlapping ranges of K^*, making the comparison difficult. Hence it was decided to use the model results for such a comparison. Fortunately, because the validity of these models has been confirmed by experimental measurements, the present comparison is considered meaningful. In it, we use the collection efficiencies calculated for 95% relative humidity. These are taken from the results obtained in Wang *et al.* (1978) for drops, Martin *et al.* (1980a) for hexagonal ice plates, and Miller and Wang (1989) for columnar ice crystals. Note that the drop collection efficiencies are calculated for $P = 1000$ mbar, and $T = 20°C$, while the ambient conditions for the ice crystals are $P = 700$ mbar and $T = -10°C$ for plates versus $P = 600$ mbar and $T = -20°C$ for columnar ice crystals. Wang *et al.* (1978) have shown that the pressure and temperature conditions have little effect on the collection efficiency. Note also that we define the relative humidity (RH) here as that with respect to saturation vapor pressures over plane water and ice surfaces instead of the commonly used definition, which uses saturation over plane water surface as the standard. Thus, the condition RH = 95% is different for drops than for ice particles. But again, the resulting differences between collection efficiencies using different definitions are typically only a few percent, and hence would not alter the general conclusions discussed below.

5.7.4. Comparison of the Aerosol Collection Efficiencies of Water Drops and Ice Crystals

Figures 5.16 through 5.18 show comparisons of the aerosol collection efficiencies by water drops and ice crystals. We shall assume that the retention efficiency is unity, so that the collision efficiencies are the same as the collection efficiencies.

Figures 5.16 and 5.17 show comparative expressions for K^* appropriate to a drop and versus a columnar ice crystal. We note that over the whole range of collector sizes considered, and for both aerosol particle sizes studied, drops are better aerosol particle scavengers than are columnar ice crystals; i.e., for a given combination of relative humidity and for a given K^*, the collision efficiency of a water drop is larger than that of a columnar ice crystal.

The situation for ice plates is somewhat different. A comparison of E as a function of K^* for a spherical drop versus a platelike ice crystal is given in Figure 5.18. We note that for all except the smallest ice crystals, plates are better scavengers of aerosol particles than are drops. This result must be attributed to the ice crystal rim, which acts as an efficient aerosol particle "trap." With decreasing values of K^*, i.e., of crystal size, the difference between collision efficiency for ice plates versus water drops decreases quickly until, for sufficiently small plates (or values of K^*), water drops dominate ice crystal plates as particle scavengers. This behavior is understandable since, with decreasing Reynolds number, the flow field around a crystal plate differs less and less from that around a sphere.

The above figures are limited to a particular aerosol particle size. A more comprehensive comparison would be one that compares the efficiencies over the entire range of the aerosol particle sizes r_p, as is done below.

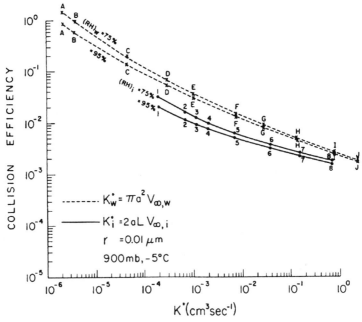

FIG. 5.16. Efficiency with which columnar ice crystals and water drops collide with aerosol particles of $r = 0.01$ μm as a function of the parameter K^*.

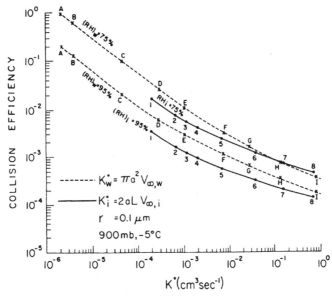

FIG. 5.17. Efficiency with which columnar ice crystals and water drops collide with aerosol particles of $r = 0.1$ μm as a function of the parameter K^*.

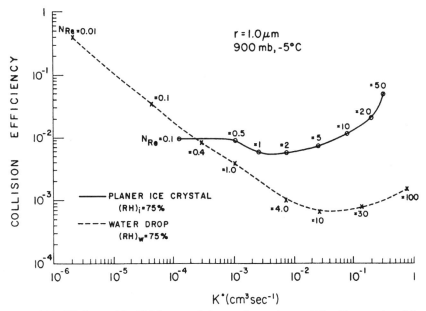

FIG. 5.18. Efficiency with which ice crystal plates and water drops collide with aerosol particles of $r = 1.0$ μm as a function of the parameter K^*.

Figures 5.19 and 5.20 show the collection efficiencies of water drops, hexagonal plates, and columnar ice crystals as functions of K^* and particle radius r_p. In order to emphasize the difference among these efficiencies, the logarithmic scales are used. The ranges for the original values are from 10^{-6} to 1 for E, from 0.001 to 10 μm for r_p, and from 3.7×10^{-4} to 0.6 cm^3 s^{-1} for K^*. These two figures are based on the same data sets except that the plots are viewed from different angles. It is immediately seen from Figure 5.19 that the collection efficiencies are very similar to each other for all three collector types. This indicates that these three kinds of collectors are almost equally efficient in collecting aerosol particles of radius about 0.001 μm. The reason for this is that, in this particle size range, the most important collection mechanism is aerosol Brownian diffusion, which depends mainly on the aerosol particle size. The efficiencies are therefore fairly insensitive to the habit of the collector. The collection efficiencies also decrease with increasing K^* for all three collectors at about the same rate. This is entirely due to the way efficiency is defined in the first equation of Section 5.7.2. Here the collection kernel K is virtually the same, but E decreases as K^* increases. Thus, as all three collectors become larger, their efficiencies decrease because the number of particles they can capture remains practically the same even though the volumes swept by them per unit time increase. Subtle differences exist among the three collectors, but the magnitudes are insignificant. Moving along the r_p axis

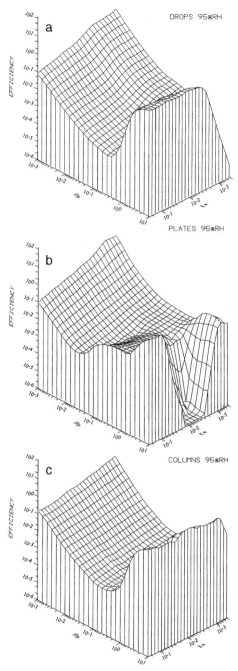

FIG. 5.19. Collection efficiencies of aerosol particles captured by (a) water drops, (b) ice plates, and (c) ice columns at RH = 95% as computed by Wang *et al.* (1978), Martin *et al.* (1980a), and Miller and Wang (1989). The particle radius R_p is in microns and the unit for K^* is cm^3 s^{-1}.

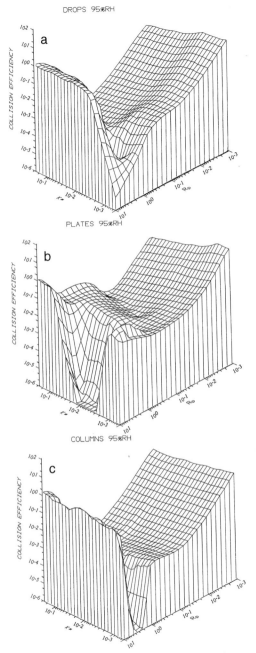

FIG. 5.20. Collection efficiencies of aerosol particles captured by (a) water drops, (b) ice plates, and (c) ice columns at RH = 95% as computed by Wang *et al.* (1978), Martin *et al.* (1980a), and Miller and Wang (1989). The particle radius R_p is in microns and the unit for K^* is cm^3 s^{-1}. (These figures are the same as those in Fig. 5.19 except viewed from a different angle.)

reveals more significant differences between the three collectors. If we look at E for larger K^* (toward the left, along the K^* axis), we see that the efficiencies for drops and columns are similar but those for plates are drastically different. First, E for all three collectors initially decreases with increasing r_p due to the decrease in Brownian collection. For drops and columns, however, the efficiencies reach a minimum near $r_p = 1$ μm, then increase with r_p as inertial impaction becomes important. For plates, the minimum E occurs for r_p between 0.01 and 0.1 μm. It is also obvious that E for plates for r_p between 0.02 and a few microns is much higher than for the drops and columns. For example, at $r_p = 1$ μm, E for plates is typically at least two orders of magnitude higher than for columns and drops. This size is in the so-called Greenfield gap range, so that plates are much more efficient in removing the Greenfield gap particles than columns and drops are. Figure 5.19 reveals E only for larger K^* values. To see the other side of the story, therefore, we have to look at Figure 5.20, where we see that what we said in the previous paragraph is also true for smaller K^* values. Unlike for the drops and columns, the E surface for plates does not have a very deep gap along the r_p axis. The gap seems to be especially deep (and narrow) for columns. For drops, the gap is more gradual, although there is a very deep "pit" near $K^* = 2.7 \times 10^{-3}$, a value which corresponds to drop radius of 42 μm, which is not very easy to see in this type of diagram. The gaps for both columns and droplets are much deeper on this side of the E surface, where the collector sizes are smaller, than on the other side. Thus, small columns and droplets are hardly effective in removing aerosol particles of about 1 μm radius. Before reaching this gap region, all three collectors have relatively smooth E surfaces, indicating that the variation of E with K^* is gradual. This remains the case for columns and drops throughout the plotted range of K^* values. But the situation for plates is again very different. We see from both Figures 5.19 and 5.20 that there is a deep "crack" in E for plates along the K^* axis. This crack encompasses plate radii between about 100 and 200 μm. The efficiencies are rather small for these plates at $r_p > 1$; indeed, if we examine the E curves in Martin et al. (1980a,b), we see that there are local maxima of E at $r_p \approx 1$ μm for $N_{Re} = 1, 2,$ and 5. The efficiency decreases with increasing r_p for these cases, as can also be seen more clearly in Figure 5.20 along the r_p axis. This special collecting behavior of plates is probably due to the strong pressure buildup near the stagnation point. Unlike columns and drops, whose surface curvatures near the stagnation points remain constant, those of plates are fairly flat. This results in high-pressure buildup over a larger area on the plate surface than on drops and columns. This large high-pressure region prevents easy capture of aerosol particles, which may thus get deflected away from the surface, as has been pointed out by Pitter and Pruppacher (1974), Pitter (1977), and Martin et al. (1980a,b). This phenomenon is particularly effective in reducing the collection efficiency if the plates are small enough for the resulting relative velocities between them and the aerosol particles to nearly vanish, affording the particles sufficient time to be totally deflected. For smaller aerosol particles, the relative velocities will be larger, and therefore this effect is not as important.

Figure 5.19 also shows that the collection efficiencies of drops at small K^* and relatively large r_p are much smaller than those of columns and plates. This is due to the fact that at small K^*, the terminal velocities of drops are not much different from those of the aerosol particles. The collector and collectee are thus falling at about the same speed, hence the low collection efficiency. It is also seen in Figures 5.19 and 5.20 that for $r_p > 1$ μm and small K^*, columns have the highest collection efficiencies among the three collectors. At large K^*, drops have higher efficiencies than both columns and plates.

5.7.5. Conclusions

We have compared the calculated collection efficiencies of cloud droplets, ice plates, and ice columns. Several conclusions can be drawn therefrom. (1) The three collectors are about equally efficient in collecting very small ($r_p < 0.01$) aerosol particles. Thus it should be expected that the removal of these small particles should be nearly independent of the details of cloud processes, since it is mainly a function of the size of aerosol particles. Although the efficiencies of larger hydrometeors (large raindrops, graupel, hailstones) have not yet been calculated, it is to be expected that their efficiencies will not differ too much from those three smaller collectors in the removal of very small particles. Thus, to these particles, clouds and precipitation behave like filters whose filtration efficiency depends mainly on the mechanical arrangement of the filter elements (number of collectors, pressure, etc.) but not much on the physical and chemical properties of these elements. (2) Ice plates are most efficient in removing aerosol particles in the size range between 0.01 and 1.0 μm (the Greenfield gap particles). The cloud droplets and columns are generally one to two orders of magnitude less efficient than the plates. As no observations have yet been made of the habit distributions of ice crystals in clouds, it is impossible to deduce what this fact may imply. However, in light of crystal habit characteristics obtained in the laboratory experiments (see, for example, Magono and Lee, 1966; also Fig. 2.26 in Pruppacher and Klett, 1978), it may be said that plate ice crystals exist mainly in midlevels of a deep convective cloud at temperatures between -10 and $-20°$C, whereas in the higher part of the cloud at temperatures below $-20°$C, columns are dominant. Thus the Greenfield gap particles may be most efficiently removed in the midlevels of the cloud and inadequately removed in the higher levels. Thus these Greenfield gap particles may "leak" into the upper troposphere and/or lower stratosphere due to inefficient filtration of columnar crystals, whereas the middle troposphere may be "cleanest" in terms of their number concentration right after such a cloud process. This also seems consistent with the observations of Changnon and Junge (1961) where there is a local maximum of "large" particles at 15–20 km and a local minimum at 5–10 km. Of course, due to insufficient observations of both the crystal habit distributions in clouds and the actual aerosol size distributions at different levels, this remains conjecture. (3) For aerosol particles of radius greater than 1 μm, both droplets and columnar ice crystals are fairly efficient in removing them, with columns somewhat more efficient, especially when the collectors themselves are small. This may be relevant to situations near the top of a deep convective cloud just starting to glaciate. In these cases, the ice crystals are expected to be small. On the other hand, ice plates are relatively inefficient particle filters for reasons mentioned earlier. Thus, depletion of "giant" particles may be faster near the cloud top than in the middle levels. In a deep convective storm, of course, other processes must be considered before definite conclusions about the aerosol particle removal can be drawn. These processes include drop nucleation, cloud-precipitation inter-action, convection and turbulent diffusion of particle plumes, electricity, and so on. Nevertheless, it is expected that these processes can also be molded much as in the present study.

5.8. Experimental Verification of Collection Efficiencies

The collection efficiencies presented in the preceding two sections are theoretical results. In order to use them confidently to estimate the scavenging of aerosol particles by ice crystals, they have to be checked by experimental measurements. It is not necessary to produce large sets of experimental data (which is usually very difficult to do anyhow), but the data must be adequate to verify the theory. In this section we describe an experimental study of the aerosol particle scavenging by ice crystals, as performed by Sauter and Wang (1989).

5.8.1. Experimental Setup and Procedure

The experimental setup used in this study is similar to that used by Wang and Pruppacher (1977) for studying the scavenging of aerosol particles by raindrops. However, the total dimension of the present setup is smaller because snow crystals, unlike raindrops, require relatively short distances to reach terminal velocity. A schematic of the setup, including relevant dimensions, is shown in Figure 5.21.

A modified La Mer generator was used to produce indium acetylacetonate aerosol particles of radius 0.75 μm. These particles are nearly monodispersed spheres, as evidenced by electron micrographs taken after precipitation onto a glass slide by a thermal precipitator. The aerosol particles produced by the genera-tor were flushed into a Plexiglas aerosol chamber 1.21 m long and 0.155 m in inner diameter fitted with mechanical shutters at both top and bottom. This chamber was built in three pieces. In several experiments the center piece was removed, so that the chamber length was 0.61 m. When the chamber was being filled, a small air pump was run intermittently to circulate the aerosol in the chamber and ensure a uniform distribution. If the aerosol was to be electrically charged, it was first mixed with ions produced from a corona discharging unit. A mean charge of up to 26 electrons could be placed on each aerosol particle if so desired. Using a

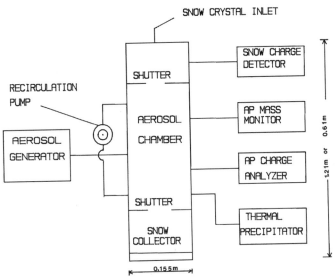

FIG. 5.21. Schematic of the experimental setup for aerosol scavenging by snow crystals.

technique developed by Dalle Valle *et al.* (1954), we determined the mean charge of aerosol particles by allowing an aerosol stream to pass through two parallel cylindrical electrodes and measuring the angle of deflection. The potential difference between the two electrodes was maintained at 10,800 V. Both the amount and the sign of aerosol charge can be determined in this way.

A typical experiment proceeded as follows. Before or during a promising snowfall, the aerosol generator was turned on. After approximately one hour, when the assembly had reached proper temperature (150°C), the aerosol particles were fed into the aerosol chamber. A TSI Model-3500 mass monitor measured the mass concentration of aerosol particles in the chamber. The particle size was measured directly under a scanning electron microscope. Typical aerosol concentration in the chamber was between 10^3 and 10^4 particles per cubic centimeter. Relative humidity in the chamber was monitored by a dewpoint hygrometer. Next, a rectangular sheet of polyethylene was coated with 2–5% (by mass) solution of Formvar and was quickly placed on a clean sheet of paper on top of an adjustable metal stand beneath the bottom aerosol chamber shutter. A hollow cylindrical piece of rigid cardboard extended approximately 20 cm below this shutter. The cardboard tube was sealed with plastic sheet and tape to keep it airtight once the stand with the Formvar-coated sheet was cranked up tight against it. At this point, both the top and bottom shutters were briefly opened. The chamber humidity was about 85% with respect to ice saturation due to some mixing of the original chamber air and the environmental air. Some loss of particles occurred during the opening, and the concentration was diluted to 30–50% of the original. Since the concentration was monitored continuously during the whole experiment, its value at the time of scavenging could be determined.

When the snowflakes fell throughout the aerosol chamber, they first passed through an induction-ring electrometer connected to an oscilloscope. A short pulse would be induced when a charged snow crystal passed through the ring. From the amplitude of the pulse, the flake charge can be evaluated. Small charges were observed (10^{-4} esu or less), in seeming agreement with the measurements of Bauer and Pitter (1982), who indicated that 50–90% of unrimed snowflakes, including those several millimeters across, have negligible charges. In addition, in an experimental study done by Magono *et al.* (1974) in Sapporo, Japan, negligible charges on natural snow crystals were measured. Most of the crystals retrieved in this study were both unrimed and small, so the small charges are probably not surprising.

After the shutters were closed, the polyethylene sheet was carefully removed and kept below freezing for at least 24 hours so that ice crystals were completely sublimed. Some crystals appeared to have melted during the collection process. As long as their size and shape could be determined, they were examined. When an appropriate snow crystal was identified, it was carefully cut out together with the surrounding polyethylene. The crystal size was measured, and the shape recorded and photographed. Originally, neutron activation analysis was employed to determine the aerosol mass on the snow crystal, but the results were unsatisfactory. It was then decided to count the particles on a snow crystal directly under a scanning electron microscope. The indium acetylacetonate particles were fairly easy to distinguish from other debris or markings on the crystal owing to their size and spherical shape. Overcounting or undercounting was not deemed to be a serious problem. Several flakes were counted twice, the second time being several days to weeks after the initial count. The dual count gave consistent results. Once the number of aerosol particles on an individual snow crystal was known, the collection efficiency was determined by the following formula:

$$E = n/(ALC) \qquad (5.34)$$

where E is the collection efficiency, n is the total number of aerosol particles on the crystal, A is the cross-sectional area of the crystal, L is the length of the aerosol chamber (either 0.61 m or 1.21 m), and C is the aerosol concentration. Collection efficiencies determined in this manner are given in the next section.

5.8.2. Results and Discussion

Seventy-two crystals were analyzed and their collection efficiencies determined. Of these, 17 were irregularly shaped and were excluded. The results for the remaining 55 crystals are given below.

Since snow crystals have complicated shapes, it is impossible to describe both the size and shape *simultaneously* by a single parameter. Therefore, in presenting the experimental results, we chose to present collection efficiencies as a function of crystal size only. For columnar crystals, we used the length as the size parameter; for planar crystals, the diameter. Other parameters such as crystal cross-sectional area, total surface area, Pasternak–Gauvin length (total surface area divided by perimeter length), and crystal Reynolds number were tried as the independent variable, but no significant difference was observed.

Figures 5.22 through 5.26 show the measured collection efficiencies versus crystal size. The data points show a fair amount of scatter, which may be due to many factors. First, the shapes of ice crystals in one category are actually not all the same. For example, of all crystals categorized as broad-branched, many have branches with different shapes and thickness. Similarly, crystals in the category of hexagonal plates may not be ideal hexagons. Some needles and columns have one end larger than the other, and some stellar crystals have branches broken off. With these nonideal size–shape situations, it is not surprising to see some scatter. Moreover, the aerosol chamber was exposed to the environmental air, which might have caused a temperature gradient so that the temperature was lower near the wall than in the center of the chamber. This could cause some nonuniformity in the aerosol concentration.

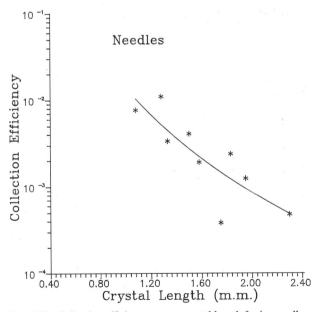

FIG. 5.22. Collection efficiency versus crystal length for ice needles.

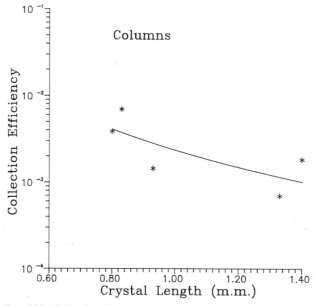

FIG. 5.23. Collection efficiency versus crystal length for ice columns.

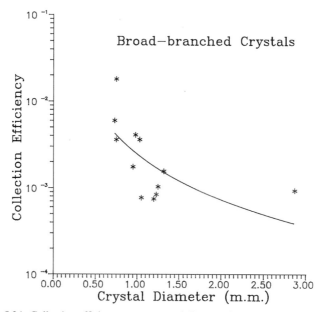

FIG. 5.24. Collection efficiency versus crystal diameter for broad-branch crystals.

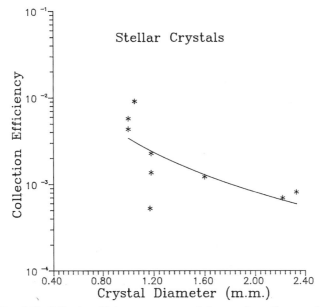

FIG. 5.25. Collection efficiency versus crystal diameter for stellar crystals.

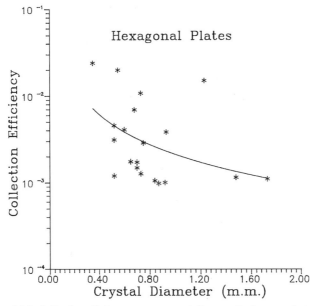

FIG. 5.26. Collection efficiency versus crystal diameter for hexagonal plates.

Despite the scatter, the results show clear trends. For a fixed crystal habit, the collection efficiency decreases with increasing collector size. Recall the definition of the collection efficiency for nonspherical collectors (Wang, 1983):

$$E = K/K^*$$

where K is the collection kernel and K^* is the geometric collection kernel, i.e., the volume swept out by the collector per unit time. For a fixed collector shape, K^* increases with increasing size. In general, K also increases with size. The decreasing trend of E indicates that the increase of K with size is less than that of K^*, consistent with what we have seen in the previous two sections. A qualitative explanation is given as follows. For the particle size considered here ($r \sim 0.75 \, \mu$m), the most important collection mechanism is inertial impaction, which depends on the relative strength of the inertial force of the particle and the hydrodynamic drag force created by the falling motion of the crystal. Since the particle size is fixed in the present study, the decrease of collection efficiency with increasing crystal size must be due to the increasing drag force. Larger crystals of the same type have higher fall speeds, but the increase in fall speed is not enough to increase the induced inertia (which scales with fall speed divided by size). Thus, the effect of particle inertia relative to particle drag decreases, causing the particle to follow the streamlines more closely. Consequently, the particle is less likely to collide with the crystal, decreasing the collection efficiency.

The foregoing discussion can be further clarified by looking at the dimensionless equation of motion of the particle:

$$dV = z/\text{Fr} - (\Delta v/V)/\text{Sk} \tag{5.35}$$

where z is the vertical unit vector, V is the vector particle velocity, and $\Delta v = V - U$ is the instantaneous particle velocity relative to the local air velocity U; Fr and Sk are the Froude number and Stokes number, respectively, defined by

$$\text{Fr} = U_\infty^2/gD \tag{5.36}$$

$$\text{Sk} = \rho_p \, d_p^2 U_\infty/18\mu D \tag{5.37}$$

where U_∞ and D are the fall speed and diameter of the crystal, g is the gravitational acceleration, μ is the dynamic viscosity of air, and ρ_p and d_p are the density and diameter of the aerosol particle. For constant particle size and environmental conditions, Sk is proportional to the ratio of the crystal terminal velocity to its dimension; i.e., $\text{Sk} \propto U_\infty/D$. Since fall speeds generally increase with the crystal size as $U \propto D^n$, where $n < 1$, then Sk decreases with increasing crystal size. For capture of particle on the upstream side of the crystal, the collision efficiency decreases monotonically with Sk.

The Froude number scales particle inertia relative to gravity, and helps explain changes in vertical forces that affect front and rear capture. For example, the increasing influence of gravity (decreasing Fr) may cause collision efficiencies to decrease on the upstream side of the collector. Changes in Fr depend on whether $n > 0.5$ or $n < 0.5$, since Fr $\propto U_\infty^{2n}/D$.

For these experiments a typical value of Sk ~ 0.02 suggests that the particle inertia is small compared to drag, whereas a typical value of Fr ~ 10 indicates that particle inertia is small compared to gravity. Even though drag dominates the equation of motion, so that particles tend to follow streamlines, inertia and gravity can still be important in the low collision efficiency problem. It is only because of such forces (and geometric interception) that the efficiency is nonzero.

It is convenient to fit the experimental data by some empirical relations. The following power equations represent the least-squares fits for the data points in Figures 5.22 through 5.26 (E = efficiency, L = crystal length in mm, D = crystal diameter in mm):

$$E = 1.42 \times 10^{-2}(L)^{-4.05}, \qquad 1.0 < L < 2.3 \text{ mm (needles)} \tag{5.38}$$

$$E = 2.35 \times 10^{-3}(L)^{-2.49}, \qquad 0.8 < L < 1.14 \text{ mm (columns)} \tag{5.39}$$

$$E = 2.45 \times 10^{-3}(D)^{-1.75}, \qquad 0.7 < D < 3.0 \text{ mm (broad-branched)} \tag{5.40}$$

$$E = 3.41 \times 10^{-3}(D)^{-2.07}, \qquad 1.0 < D < 2.4 \text{ mm (stellar)} \tag{5.41}$$

$$E = 2.12 \times 10^{-3}(D)^{-1.17}, \qquad 0.3 < D < 1.8 \text{ mm (plates)} \tag{5.42}$$

The typical values of collection efficiencies in Figures 5.22 through 5.26 are between 10^{-2} and 10^{-3}. These values are characteristic of the collection efficiency in the atmosphere for collectors of a few millimeters scavenging submicron-size aerosol particles. Such E values are also typical for drop particle scavenging (Wang and Pruppacher, 1977).

The effect of aerosol charge on the collection efficiency was also investigated. Since snow crystals in the present study had negligible charge, the only effective electrostatic force would be the image force due to the charges on the aerosol particles. The results obtained with electrically charged aerosol particles (also included in Figs. 5.22–5.26) do not show significant differences from the uncharged aerosol, indicating that the image force did not play an important role in the scavenging process. This is consistent with the finding of Wang and Pruppacher (1980a), who concluded that image force is unimportant in the scavenging of aerosol particles by water drops in the presence of an external electric field.

Note that the insignificance of the electric effect in the present study is due to the negligible charges on the snow crystals. In other situations where snow

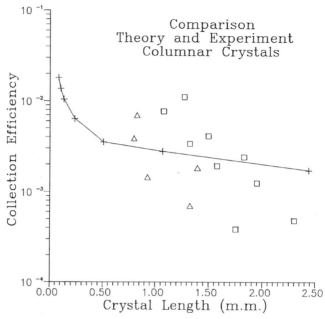

FIG. 5.27. Comparison of theoretical and experimental collection efficiencies for aerosol particles captured by columnar ice crystals. Solid curve, theoretical predictions; △, needles (experimental); □, column (experimental).

crystals may carry substantial charges, the electric effect may become important. We have inadequate knowledge of the electric charges on ice crystals in thunderclouds and snowstorms. If these crystals are significantly charged, the electric effect may be expected to play an important role in the scavenging of particles in these clouds.

For columnar ice crystals (including needles), Figure 5.27 shows the comparison between the experimental results and the collection efficiencies predicted by the theoretical method described in Section 5.6 under the same experimental and atmospheric conditions. It is understood here that due to the many experimental difficulties and nonideal conditions (for example, the crystals are rarely exactly columnar), an exact comparison is not possible. The only parameter that can be used as the basis of comparison is the crystal length. Nevertheless, comparison of theory to experiment, for this single particle size and atmospheric environment, shows that the two are in the same general magnitude range. Moreover, both show a trend of decreasing efficiency with increasing crystal dimension.

6. EVOLUTION OF ICE CRYSTALS IN THE DEVELOPMENT OF CIRRUS CLOUDS

6.1. Cirrus Clouds, Radiation, and Climate

It is well known that cirrus clouds are composed almost completely of ice crystals. The common perception of cirrus is that they are thin clouds, but in reality they can be rather thick, sometimes reaching a vertical extent of a few kilometers (Fig. 6.1). Possibly, they appear thin because substantial parts of them are subvisual.

In spite of their tenuous appearance, cirrus clouds have a pronounced influence on climate, owing to their effect on the radiation (Ramanathan *et al.*, 1983). Cirrus clouds are usually located high in the troposphere where temperatures are low. By

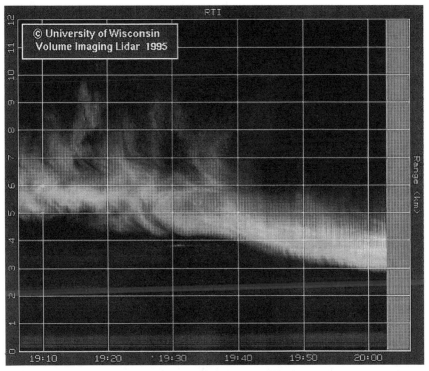

FIG. 6.1. The vertical profile of a cirrus cloud as a function of time as seen by a vertically pointing lidar. In a quasi–steady state condition, this profile is nearly equivalent to a snapshot of the cirrus at a certain time. (Courtesy of Dr. Edwin Eloranta.)

virtue of their low temperature, they will interact strongly with the upwelling and downwelling infrared radiation in the atmosphere, as dictated by the Kirchhoff law. In addition, satellite observations indicate that cirrus clouds cover extensive areas of the earth (Warren *et al.*, 1986, 1988; Wylie and Menzel, 1989). Together, these two factors imply that cirrus clouds can influence the radiative budget of the earth–atmosphere system significantly. Their radiative effects can be highly variable given the high variability in their radiative and microphysical properties. Either cooling or warming can occur, depending on the cloud radiative properties, cloud height, and the clouds' thermal contrast with the surface (Manabe and Strickler, 1964; Cox, 1971).

Randall *et al.* (1989), via a general circulation model (GCM), showed that upper tropospheric clouds have dramatic impacts on the large-scale circulation in the tropics, with attendant effects on precipitation and water vapor amounts. Ramaswamy and Ramanathan (1989), also through GCM studies, suggested that the discrepancies between previous simulations and observed upper tropospheric temperature structure in the tropics and subtropics can be explained by the radiative heating effects of cirrus cloud systems. These studies point out that cirrus clouds are likely to have great impacts on the radiation and hence the intensity of the general circulation.

The cloud forcing in GCMs is currently a major uncertainty factor. Cess *et al.* (1989), who compared the outputs of 14 GCMs simulating an equivalent climate change scenario, found that the results of global temperature change in response to an imposed sea surface temperature change were relatively uniform when clear sky conditions were assumed for radiative computations. The results were very different when the radiative effects of clouds were included. They also found that the effect of cloud feedback was comparable in magnitude to that due to imposed forcing, i.e., the change in sea surface temperature, but the sign can be positive or negative depending on the model chosen. Needless to say, this does not build confidence in the model predictions, and there is an urgent need to reduce the uncertainty in the cloud radiative forcing in GCMs and climate models.

The radiative properties of cirrus clouds depend on their microphysical characteristics such as ice crystal size, concentration, habit, and spatial distribution. The uncertainty about the radiative properties of cirrus comes from our inadequate understanding of their microphysical behavior. One way to improve this understanding is to perform model studies, provided, of course, that the model is adequately realistic.

In the following, a recent model study by our research group on the evolution of ice crystal microphysics in cirrus clouds is described. The details of the physics and mathematics of the model can be found in Liu (1999) and Liu *et al.* (2001a,b,c). Here, we describe the essence of the model briefly, but the major results in some detail.

6.2. Physics of the Model

The basic ideas of the cirrus model used for the present study were derived from an earlier work by Starr and Cox (1985a,b), but the details differ significantly. In this section, the main elements of the model is discussed.

6.2.1. Model Philosophy and Components

Three types of physical processes that have been identified as essential for the development of cirrus clouds are dynamic, microphysical, and radiative. Cirrus clouds often form during the large-scale lifting of moist air, starting as small ice crystals. If the upward motion persists long enough to cause further cooling of the layer, ice crystals will grow to sizes with substantial fall velocities; i.e., precipitation will occur. As ice crystals grow larger, the radiative effect becomes more significant. The resulting radiative heating profile changes the temperature lapse rate, and with it the dynamics in the cloud. A change in cloud dynamics will affect the microphysical processes, altering the size distribution of ice crystals. This, in turn, further modifies the radiative heating profiles within the cloud. It is clear that these processes are interactive, as shown in Figure 6.2. None of the three processes should be ignored in developing a cirrus model.

The dynamics model is a modified 2-D version of the dynamics framework used in the 3-D WISCDYMM described by Straka (1989) and Johnson et al. (1993, 1994). The main modification is in the advection scheme. For the turbulent kinetic energy, water vapor, and potential temperature, we use the sixth-order Crowley scheme (see Tremback et al., 1987). The numerical method used to calculate the advection of hydrometeors, however, is the total variation-diminishing (TVD) scheme as described by Yee (1987). The TVD scheme is introduced here because most other numerical advection scheme are dispersive across the discontinuity (the interface between the environment and the advected property), and thus may lead to negative values, which are unphysical for positive definite variables such as mixing ratios and concentrations of hydrometeors. TVD schemes, in contrast, can effectively eliminate the dispersive oscillation across the discontinuity.

In the microphysical module, a double-moment scheme is used to predict the evolution of the ice crystal size distribution at each grid point. Both the mixing ratio and the number concentration of ice crystals are prognostic variables, from which the distribution mean diameter is then diagnosed. This is more realistic than predicting mixing ratio only, as most models with bulk microphysics do. The growth rate of an ice crystal is explicitly calculated in this model. In the growth equation of ice crystals, both the capacitance and the ventilation coefficients are functions of ice crystal shape. Ventilation coefficients for different shapes of ice crystals commonly observed in cirrus are computed using the method outlined in Section 4.3. Another important microphysical process, homogeneous freezing

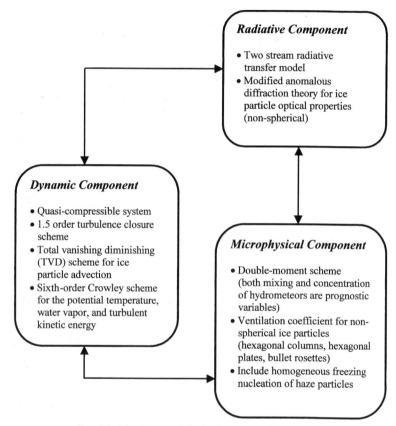

FIG. 6.2. The three modules in the present cirrus model.

nucleation, is included in our model since this has been recognized as a very effective source of ice crystals in cirrus.

Because cirrus cloud decks are usually optically thin and the mean free path for a photon colliding with a particle in cirrus is much larger than in a typical stratocumulus, the radiative heating is distributed through the entire cirrus cloud body instead of being distributed like two Dirac functions with opposite signs at the cloud top and bottom as in a typical stratocumulus cloud (Ackerman *et al.,* 1988). Moreover, the volume absorption coefficient and the volume extinction coefficient are very sensitive to the ice crystal size distribution. As the cloud evolves, the change in ice crystal size distribution causes changes in the radiative heating rates not only within, but also below and above the cloud deck. It is therefore important to represent the ice crystal optical properties correctly. For this purpose, a modified anomalous diffraction theory (MADT) proposed by Mitchell (1996) is employed. In comparison to most of the existing ice crystal optics parameterizations that are derived from limited sets of observations or laboratory results, MADT has more physics. Its parameterization is based on the physics of how the incident ray interacts with a particle. Through anomalous diffraction theory, analytical expressions are developed describing the absorption and extinction coefficients and the single scattering albedo as functions of size distribution parameter, ice crystal shape, wavelength, and refractive index. Therefore, the optical properties calculated are not based on an effective radius that has little physical meaning. Another advantage of MADT is that the scattering properties in the thermal infrared spectral range can be explicitly calculated, so that the scattering is not ignored. The radiative fluxes are calculated using a two-stream model. More details of the cloud microphysics and radiative modules are described in the following sections.

6.2.2. The Microphysics Model

As mentioned earlier, a detailed double-moment parameterization scheme is used in this model. This scheme assumes that various ice categories may be represented by continuous size distribution functions. Parameterizations are then developed for various physical processes including nucleation, diffusional growth, and collisional growth based on the assumed size distributions. These parameterizations determine how mass is transferred between various hydrometeor categories. Realistic evolution of the ice crystal size distribution is promoted via a double-moment parameterization as noted earlier.

Figure 6.3 shows the schematic of the cloud microphysical processes included in this model. Three categories of hydrometeors are considered: haze particles, pristine ice crystals, and aggregates. The size spectra of these hydrometeors are assumed to follow the inverse exponential distribution analogous to the well-known Marshall–Palmer distribution for raindrops (Marshall and Palmer, 1948):

$$F(D) = \frac{1}{D_n} \exp\left(-\frac{D}{D_n}\right) \tag{6.1}$$

where $F(D)$ is the probability distribution function defined over a size interval, D is the diameter of the ice crystal, and D_n is the mean diameter. The number density, $n(D)$, is determined by

$$n(D) = N_t F(D) \tag{6.2}$$

where N_t is the total number concentration of the ice crystal species. Ice crystals are assumed to satisfy certain characteristic dimensional relationships as described in Auer and Veal (1970), Heymsfield (1972), and Mitchell et al. (1990). These relationships are expressed in terms of power laws and are given in Table 6.1.

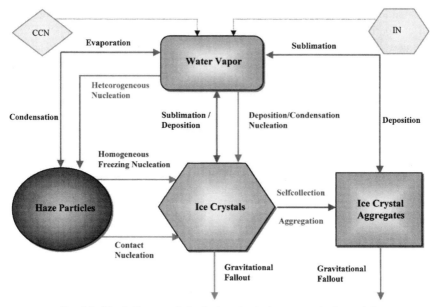

FIG. 6.3. Block diagram of cloud microphysical processes in the model.

Starr and Cox (1985b) have shown that the evolution of cirrus clouds is very sensitive to the terminal velocities of the ice crystals because they determine the sedimentation rate and hence the vertical extent and optical depth of the cloud. In this model we used a method similar to that of Böhm (1989) and Mitchell (1996) to determine the terminal velocities of ice crystals.

As indicated in Figure 6.3, the homogeneous freezing nucleation, heterogeneous nucleation, diffusional growth and sublimation, and aggregation of ice crystals are all considered in this model. The homogeneous freezing parameterization is

TABLE 6.1 POWER LAW RELATIONSHIPS FOR ICE CRYSTALS (c.g.s. UNITS)

Habit	Dimensional relationship	Projected area-dimensional relationship	Mass-dimensional relationship
Spheres	$W = D$	$A = (\pi/4)D^2$	$m = \rho_i(\pi/6)D^3$
Plates	$W = 0.0141D^{0.475}$	$A = 0.2395D^{1.855}$	$m = 0.007384D^{2.449}$
Columns	$D = 0.26L^{.927}$	$A = 0.0459L^{1.415}$	$m = 0.01658L^{1.91}$
Rosettes	$W = D$	$A = 0.0869D^{1.57}$	$m = 0.0459D^{1.415}$
Aggregates	$W = D$	$A = 0.2285D^{1.88}$	$m = 0.00281D^{2.1}$

TABLE 6.2 ICE CRYSTAL CAPACITANCE

	Expression	Sources
Columns	Approximated by charger circular cylinders: $C = [0.708 + 0.615(b/a)^{0.76}]a$	Smythe, 1956, 1962; Wang, 1985
Plates	Approximated by oblate spheroids: $C = a\varepsilon/(\sin^{-1}\varepsilon), \quad \varepsilon = [1 - (b^2/a^2)]^{0.5}$	Pruppacher and Klett, 1978
Spheres/rosettes	$C = \frac{1}{2}D$	

a = semi-major axis length; b = semi-minor axis length; D = diameter of a sphere.

taken from DeMott (1994). The parameterization of Meyers *et al.* (1992) is used to determine the heterogeneous nucleation rate due to deposition and embedded freezing (contact nucleation, on the other hand, is not considered).

The diffusional growth rate of ice crystals is calculated based on the approximation of Srivastava and Coen (1992). The saturation vapor pressure over the ice surface, a necessary quantity, is determined by using the eighth-order polynomial fit of Flatau *et al.* (1992).

The capacitances of ice crystals are also needed to determine their diffusional growth rates. The capacitances of ice columns are approximated by those of conducting finite cylinders as given in Smythe (1956, 1962) and Wang *et al.* (1985). The capacitances of ice plates are approximated by those of conducting oblate spheroids as described in Pruppacher and Klett (1997). As for rosettes, both McDonald (1963) and Heymsfield (1975) pointed out that the capacitance for particles, though with more intricate pores in them, could be approximated by that for spherical particles, as is done in this study. The capacitance formulas mentioned above are given in Table 6.2. For the convenience of computation, these formulas are approximated by power laws as shown in Table 6.3.

When particles are large enough to have sustained terminal velocities, the ventilation effect on the growth rate has to be considered. For ice columns and plates, the ventilation coefficients are taken from Ji and Wang (1998; see Section 4.3).

TABLE 6.3 COEFFICIENTS AND POWERS FOR ICE
 CRYSTAL CAPACITANCE (m)

Ice crystal type	Power law
Columns	$0.278D^{0.97}$
Plates	$0.277D^{0.99}$
Spheres/rosettes	$0.5D$
Columns	$0.278D^{0.97}$
Plates	$0.277D^{0.99}$
Spheres/rosettes	$0.5D$

For ice spheres, the values are taken from Hall and Pruppacher (1976). Note that these are just the ventilation coefficients for individual ice crystals. In the model we need to compute the distribution mean ventilation coefficient across the whole size spectrum of ice crystals so that the mean mass diffusional growth rate can be determined. This distribution mean ventilation coefficient is

$$\bar{f} = \int_0^\infty C f_v F(D)\, dD \tag{6.3}$$

where C is the capacitance and f_v is the ventilation coefficient of the ice crystal.

The aggregation of ice crystals also contributes to the change in their size distribution. The growth rate due to aggregation is determined by solving the stochastic collection equation:

$$\left.\frac{dm}{dt}\right|_{xy} = \frac{\pi N_x N_y}{4\rho_0} \int_0^\infty \int_0^{D_c} E_{x,y} m_x(D_x)(D_x + D_y)^2$$

$$\times |v_x(D_x) - v_y(D_y)| f_x(D_x) f_y(D_y)\, dD_x\, dD_y \tag{6.4}$$

and the change in concentration collected by species x is

$$\left.\frac{dN}{dt}\right|_{xy} = \frac{\pi N_x N_y}{4} \int_0^\infty \int_0^{D_c} E_{x,y}(D_x + D_y)^2 |v_x(D_x) - v_y(D_y)|$$

$$\times f_x(D_x) f_y(D_y)\, dD_x\, dD_y \tag{6.5}$$

Here $E_{x,y}$ is the collection efficiency, which is the product of the collision and coalescence efficiencies; D_c is the diameter of species x, whose volume V_c is equal to or less than that of species y. Note that in Eqs. (6.4) and (6.5) the smaller particle can have a greater terminal velocity than the larger particle. In the above equations, only particles of species x are allowed to be collected by species y. This is done so as to prevent duplicate collections which occur when species x collects species y. Since the stochastic collection equation has no straightforward analytical solution, solutions for Eqs. (6.4) and (6.5) are calculated by numerically integrating the equation over 100 discrete size intervals in D_x and D_y.

There are analytical solutions for Eqs. (6.4) and (6.5) for collection within the same ice category (self-collection process). According to Verlinde et al. (1990), the analytical solutions for the change in mixing ratio due to self-collection can be written as

$$\left.\frac{dq}{dt}\right|_{self} = \frac{1}{\rho_0} \frac{\pi}{4} E J_{xx} \tag{6.6}$$

where

$$J_{xx} = \tfrac{1}{2}N_t^2 m(D_n)V_t(D_n)D_n^2 C_{xx} \tag{6.7}$$

$$
\begin{aligned}
C_{xx} = \sum_{n=0}^{2} &\left[\frac{2}{v+n}\Gamma(\eta)_2 F_1(v+n, \eta; v+n+1; -1) \right.\\
&\left. - \Gamma(v+n)\Gamma(p_m + p_v + v - n + 2) \right]\\
+ \sum_{n=0}^{2} &\left[\frac{2}{v+p_v+n}\Gamma(\eta)_2 F_1(v+p_v+n, \eta; v+p_v+n+1; -1) \right.\\
&\left. - \Gamma(v+p_v+n)\Gamma(p_m + v - n + 2) \right]
\end{aligned} \tag{6.8}
$$

and

$$\eta = p_m + p_v + 2v + 2 \tag{6.9}$$

The change in number concentration due to self-collection can be expressed as

$$\left. \frac{dN}{dt} \right|_{self} = \frac{\pi}{4}E J_{xx} \tag{6.10}$$

where

$$J_{xx} = \tfrac{1}{2}N_t^2 V_t(D_n)D_n^2 C_{xx} \tag{6.11}$$

$$
\begin{aligned}
C_{xx} = \sum_{n=0}^{2} &\left[\frac{2}{v+n}\Gamma(\eta)_2 F_1(v+n, \eta; v+n+1; -1) \right.\\
&\left. - \Gamma(v+n)\Gamma(p_v + v - n + 2) \right]\\
+ \sum_{n=0}^{2} &\left[\frac{2}{v+p_v+n}\Gamma(\eta)_2 F_1(v+p_v+n, \eta; v+p_v+n+1; -1) \right.\\
&\left. - \Gamma(v+p_v+n)\Gamma(v - n + 2) \right]
\end{aligned} \tag{6.12}
$$

and

$$\eta = p_v + 2v + 2 \tag{6.13}$$

In the above equations, $\Gamma_2 F_1$ is the Gaussian hypergeometric function, p_v is the exponent in the power-law relationship for terminal velocity, p_m is the exponent

TABLE 6.4 BANDS IN THE RADIATIVE TRANSFER MODULE (SOLAR SPECTRUM)

Band number	Band center (μm)	Fractional solar energy distribution (%)	Absorption coefficient K_N	Weighting function W_N	Gaseous absorption
1	0.25	1.2019	135.000000000*	0.68594100000	O_3
			12.549720000	0.29036870000	
			0.000000000	0.00003220838	
2	0.35	7.5150	6.091203000*	0.12196820000	O_3
			1.177907000	0.07025021000	
			0.572324000	0.23763020000	
			0.000000000	0.57011060000	
3	0.45	13.8740	0.153407000*	0.02926001000	O_3
			0.115351000	0.01340009000	
			0.000000000	0.95734590000	
4	0.55	13.0840	0.115351300*	0.47060510000	O_3
			0.056013310	0.46009560000	
			0.000000000	0.06927936000	
5	0.65	11.1960	0.115351300*	0.25875810000	O_3
			0.056013310	0.74121330000	
6	0.78	13.7430	60.857640000*	0.00003323726	O_3
			0.007011613	0.69653730000	
			0.003504144	0.30343740000	
7	0.94	11.4830	135.000000000*	0.00432769880	H_2O
			36.586678000	0.00774681680	
			8.313803000	0.01625321900	
			2.362969000	0.02797893000	
			1.529817100	0.01446416500	
			0.563513920	0.07340810400	
			0.089909350	0.21243957000	
			0.000000000	0.64278851000	
8	1.14	6.2990	135.000000000*	0.00880807980	H_2O
			52.704774000	0.00123414800	
			17.572670000	0.01761926000	
			4.598316400	0.02863373200	
			1.497370000	0.04850655200	
			0.464925900	0.08106161100	
			0.087403004	0.14613502000	
			0.000000000	0.66694856000	
9	1.38	10.2070	135.000000000*	0.05729474000	H_2O
			45.155164000	0.00990807900	
			15.270085000	0.10932248000	
			3.191396700	0.11710367000	
			0.697789060	0.10823056000	
			0.114325470	0.14210582000	
			0.000000000	0.44322885000	
			0.000000000	0.00000000000	

TABLE 6.4 (*continued*)

Band number	Band center (μm)	Fractional solar energy distribution (%)	Absorption coefficient K_N	Weighting function W_N	Gaseous absorption
10	1.87	6.7630	135.000000000*	0.04863024000	H_2O
			45.155160000	0.01039200000	
			15.270090000	0.08680198000	
			3.258179000	0.08230453000	
			0.666104600	0.06686005000	
			0.079500980	0.17443730000	
			0.000000000	0.51901520000	
			0.000000000	0.00000000000	
11	2.70	3.6100	135.000000000*	0.18767610000	H_2O
			45.155160000	0.08640150000	
			15.221560000	0.06783896000	
			8.911640000	0.14627670000	
			1.872147000	0.15094330000	
			0.400341300	0.20267700000	
			0.048934550	0.10063670000	

*Denotes absorption coefficients for Rayleigh absorption.

in the power-law relationship for ice crystal mass, and the rest of the variables are as previously defined. The collection efficiency, which is used in the collection equations, can be factored into two components, the collision efficiency (E_{coll}), and the coalescence efficiency (E_{coal}):

$$E = E_{coll} \times E_{coal} \qquad (6.14)$$

In this study, the coalescence efficiency is assumed to be 1, and the collision efficiency is set at 0.1 (Kajikawa and Heymsfield, 1989).

6.2.3. The Radiative Transfer Model

In order to calculate the radiative heating rates, the radiative transfer equations must be solved and the cloud optical properties determined. Since the upward and downward fluxes at a given layer in the atmosphere are the main concern here, it is not necessary to calculate the radiance distribution at each level. Instead, a two-stream/adding model (Ackerman and Stephens, 1987) is used to calculate radiative fluxes at each grid point. The one applied here is a narrow-band model, which divides the solar and terrestrial radiation spectrum into 11 and 20 spectral bands, respectively. Tables 6.4 and 6.5 summarize these bands. The cloud optical properties are computed using the modified anomalous diffraction theory (MADT) proposed by Mitchell (1996, 1998) for ice particles of different shapes. With the use of MADT, analytical expressions can be obtained for describing the absorption and extinction coefficients and the single scattering albedo as functions of size

PAO K. WANG

TABLE 6.5 BANDS IN THE RADIATIVE TRANSFER MODULE (TERRESTRIAL SPECTRUM)

Band number	Band center (μm)	Band width (μm)	Absorption coefficient K_N	Weighting function W_N	Gaseous absorption
1	4.71	3.00	1.02270	0.24189	H_2O
			0.09935	0.64075	
			7.46580	0.11349	
2	5.00	4.99	0.24615	0.37737	H_2O
			1.24530	0.29677	
			4.70600	0.16017	
			77.18300	0.06306	
			17.33800	0.10260	
3	5.26	5.13	1.02390	0.23925	H_2O
			4.61050	0.37021	
			47954.00000	0.00853	
			21321.00000	0.10218	
			26.25500	0.27983	
4	5.56	5.41	49.70900	0.33730	H_2O
			217.62000	0.33385	
			2844.10000	0.10910	
			41242.00000	0.00455	
			807.17000	0.21519	
5	5.88	5.71	44.54700	0.33757	H_2O
			200.56000	0.35053	
			2079.00000	0.08749	
			38140.00000	0.00135	
			682.94000	0.22305	
6	6.25	6.06	27.79500	0.41228	H_2O
			142.99000	0.30772	
			1209.88000	0.09035	
			3390.50000	0.02281	
			487.11000	0.16583	
7	6.67	6.45	12.40400	0.37385	H_2O
			58.74100	0.32079	
			496.14000	0.09805	
			1247.70000	0.02305	
			168.56000	0.18425	
8	7.14	6.90	8.38210	0.22777	H_2O
			2.95740	0.07143	
			24.47500	0.26819	
			347.17000	0.02050	
			79.97900	0.22761	
9	7.70	7.40	3.34820	0.25107	H_2O
			24.77400	0.19668	
			0.75039	0.33681	
			0.18471	0.21535	
10	8.70	8.00	1.00000	1.00000	H_2O

TABLE 6.5 (*continued*)

Band number	Band center (μm)	Band width (μm)	Absorption coefficient K_N	Weighting function W_N	Gaseous absorption
11	9.70	9.40	119.76600	0.06310	H_2O, O_3
			22.26470	0.12344	
			2647.50000	0.63267	
			64387.00000	0.18079	
12	10.50	10.00	1.00000	1.00000	H_2O
13	11.75	11.00	1.00000	1.00000	H_2O
14	12.90	12.50	135.30000	0.01493	H_2O
			3.75265	0.05505	
			0.53660	0.10154	
			0.08288	0.49848	
			0.00000	0.32999	
15	15.30	13.30	135.30000	0.05973	H_2O, CO_2
			3.75265	0.13612	
			0.64362	0.26775	
			0.10837	0.41955	
			0.00000	0.11683	
16	18.62	17.24	135.00000	0.13599	H_2O
			10.66490	0.15818	
			3.75716	0.19582	
			1.37011	0.23766	
			0.47950	0.23491	
			0.18251	0.03807	
17	22.50	20.00	135.00000	0.22419	H_2O
			12.91897	0.32042	
			6.01122	0.06852	
			2.33170	0.32235	
			0.70562	0.06242	
18	28.60	33.30	135.00000	0.75308	H_2O
			5.80000	0.24472	
			0.00000	0.00213	
19	40.00	50.00	61.06500	0.09530	H_2O
			554.22000	0.25561	
			9745.10000	0.20390	
			1884.40000	0.19568	
			182.85000	0.24952	
20	100.00	125.00	389.43000	0.23989	H_2O
			17814.0000	0.19969	
			1159.4000	0.26330	
			3966.1000	0.21371	
			133.6000	0.08337	

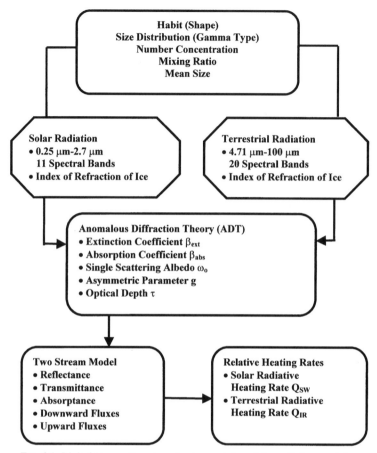

FIG. 6.4. Links between the microphysics module and the radiation module.

distribution parameter, ice crystal shape, wavelength, and refractive index. The links between the radiative transfer module and the microphysics module to calculate the radiative heating rates are illustrated in Figure 6.4. By using this method, the absorption coefficient, extinction coefficient, and single scattering albedo can be calculated accurately for both the solar and terrestrial spectrum and scattering is not ignored when calculating thermal infrared radiation. In this study, the radiative heating rates are calculated for the solar spectrum and terrestrial spectrum separately.

The indices of refraction for ice used in this study are from Warren (1984), and are shown in Figure 6.5. The absorption of terrestrial radiation is significant throughout the whole thermal infrared spectrum, while the absorption of solar radiation for ice is much weaker. Ice crystals are almost transparent at wavelengths less than 2 μm, so the solar heating is expected to be small compared to thermal

Real Index of Refraction of Ice

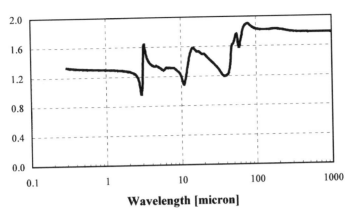

Imaginary Index of Refraction of Ice

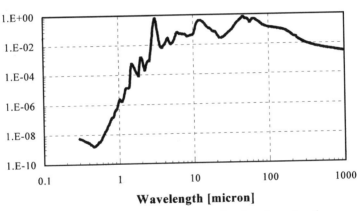

FIG. 6.5. Ice crystal refractive index as a function of wavelength.

infrared heating. On the other hand, ice has a very pronounced absorption feature near 3 μm that can lead to significant heating. In this model, vertical atmospheric columns are assumed to be independent of each other. Consequently, interactions of radiative flux among adjacent columns are ignored.

6.3. Design of the Present Simulation Study

There is a wide range of possible environmental and physical factors that may influence the development of cirrus clouds. Table 6.6 lists several physical parameters that are important to cirrus clouds. Environmental forcing such as large-scale lifting (vertical motion) and static stability are important, as are cloud height

TABLE 6.6 IMPORTANT ENVIRONMENTAL AND MICROPHYSICAL FACTORS FOR CIRRUS CLOUDS

Parameters	Combinations
Cloud height (temperature)	Warm versus cold cirrus
Large scale forcing	Weak versus strong vertical uplifting
	Cirrus uncinus: 100–150 m/s
	Warm front overrunning: 2–10 cm/s
	Warm front occlusion: 20 cm/s
	Cloud low aloft: 25–50 cm/s
Horizontal wind shear	No shear versus weak to strong shear (usually positive)
	Ranging from about 4 m s^{-1} km^{-1} in summer to 6.5 m s^{-1} km^{-1} in winter
Time of day	Nighttime versus daytime (thermal infrared only versus solar and thermal infrared radiation)
Stability (thermal stratification)	Stable condition versus conditionally unstable condition; quite variable
Ice crystal habit	Hexagonal columns
	Bullet rosettes
	Hexagonal plates
	Spheres (assumed by most models)
CCN composition and concentration	$(NH_4)_2SO_4$ versus H_2SO_4

(temperature or pressure factor), ice crystal habit, and the time of day. Another important factor, wind shear, is not considered in the present study.

To test the influence of these parameters on the evolution of ice microphysics in cirrus clouds, several model simulations are performed under different prescribed environmental and physical conditions. In particular, the following parameters are tested.

(1) Static Stability

Cirrus cloud layers are in general statically stable in an absolute sense, with lapse rates less than the moist-adiabatic rates with respect to either water or ice. Some cirrus clouds are located above stable layers, indicating that they are associated with upper-level fronts. Others are below stable layers, indicating that they are located just below the tropopause. FIRE observations revealed that cirrus could also be found in a conditionally unstable environment whose lapse rate is slightly greater than moist-adiabatic, but much less than the dry-adiabatic.

(2) Cloud Height

Another consideration for the design of the simulation sets is the cloud height, i.e., the vertical location of cirrus. Ackerman *et al.* (1988) pointed out that the cloud heating rates and cloud base warming due to thermal infrared heating are very sensitive functions of cloud height and thickness. In addition, the altitude of high cirrus clouds makes *in situ* measurements of their microphysical properties by aircraft a difficult task. A numerical simulation of such clouds may provide some useful information that is otherwise difficult to obtain.

In this study, we consider two types of atmospheric stability regimes in the simulation sets. The first is a statically stable atmosphere, and the other is a conditionally unstable layer overlying a stable layer. Each of the two stability conditions is combined with one of two upper-tropospheric cloud heights designated as "cold" and "warm," to signify a high summertime cirrus layer and lower-lying springtime cirrus layer, respectively. This forms four kinds of atmospheric background profiles, which we call cold-stable, cold-unstable, warm-stable, and warm-unstable to schematize four typical environments for cirrus. Atmospheric profiles representing these four cases were made available by cirrus observation groups at NASA (GCSS WG2), and are thought to be representative for specific seasons and locations. Table 6.7 lists some general conditions of the cold, warm, stable,

TABLE 6.7 DEFINITION OF INITIAL PROFILES

Definitions	Descriptions
WARM	U.S. Spring/Fall atmosphere
	March 21 (80 Julian day)
	Location: 45°N
	Initial supersaturated layer:
	Height: 8–9 km
	Temperature: $-37°C \sim -48°C$
	Surface temperature: 15°C
	Surface albedo: 0.2
	Tropopause at 10.5 km ($-56.6°C$)
	Simulation time: 13:00–16:00
	Solar zenith angle: 47.8°–69.9°
COLD	U.S. Summer
	June 21 (172 Julian day)
	Location: 30°N
	Initial supersaturated layer:
	Height: 13–14 km
	Temperature: $-56°C \sim -68°C$
	Surface temperature: 31.4°C
	Surface albedo: 0.2
	Tropopause at 15.5 km ($-76°C$)
	Simulation time: 13:00–16:00
	Solar zenith angle: 14.9°–53.4°
STABLE	Temperature lapse rate is 8°C/km in the initial supersaturated layer
UNSTABLE	Temperature lapse rate is ice pseudoadiabatic lapse rate for the lower 0.5 km supersaturated layer
	Temperature lapse rate is 1°C/km greater than the ice pseudoadiabatic lapse rate for the upper 0.5 km

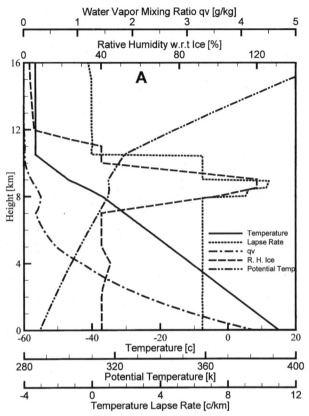

FIG. 6.6. Profiles of background atmospheric conditions for the warm-unstable case.

and unstable situations. The details of the four background profiles are given in Figures 6.6 through 6.9.

The warm cirrus profiles are based on the U.S. Standard Spring Atmosphere at 45°N. The cold cirrus profiles are based on the U.S. Standard Summer Atmosphere at 30°N. Surface temperature is 15°C for warm cirrus, and 31.4°C for cold cirrus. The corresponding background tropospheric temperature lapse rates are 6.5°/km and height-dependent. The tropopause occurs at 10.5 km (−56.5°C) and 15.5 km (−76°C), respectively. The background tropospheric relative humidity is set to 40% with respect to a plane water surface. Profiles in Figures 6.6 through 6.9, all versus height, include temperature, potential temperature, water vapor mixing ratio, temperature lapse rate, and relative humidities with respect to a plane water surface and plane ice surface. Temperature lapse rates for the warm-unstable case

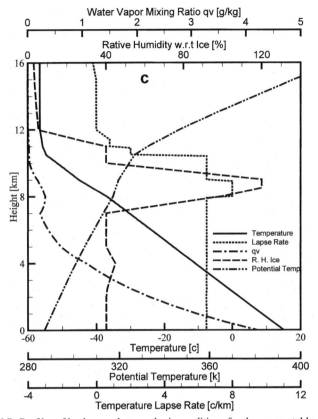

FIG. 6.7. Profiles of background atmospheric conditions for the warm-stable case.

are ice pseudoadiabatic from 8.0 to 8.5 km, and 1°C/km greater than the ice pseudoadiabatic from 8 km to 9.0 km. Similarly, temperature lapse rates for the cold-unstable case are ice pseudoadiabatic from 13 to 13.5 km, and a value 1°C/km greater than the ice pseudoadiabatic from 13.5 km to 14 km. For the statically stable cases, the lapse rates in these layers are set to 8°C/km in each profile. Relative humidity with respect to ice in the unstable cases is 100% at the base of the ice-neutral layer and increases linearly with height to the base of the conditionally unstable layer, within which it is constant at 120%. Since both daytime and night-time cirrus cases are examined in this study, some care must be taken regarding solar geometry, which is defined via a specification of latitude, date, and initial local solar time (LST). We choose March 21 (Vernal Equinox) for the warm cirrus cases and June 21 (Summer Solstice) for the cold cirrus cases. The simulations

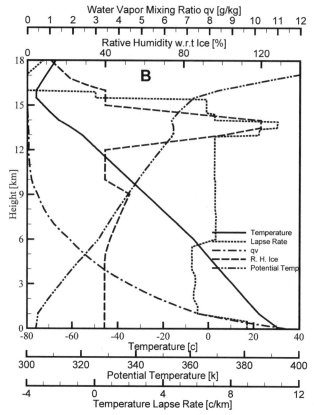

FIG. 6.8. Profiles of background atmospheric conditions for the cold-unstable case.

run for three hours from 1300 to 1600 LST. The corresponding solar zenith angle varies from 47.8° to 69.9° for the warm cirrus cases and from 14.9° to 53.4° for cold cirrus cases. We also assume that the surface albedo is 0.2, representing a climatological averaged value.

(3) Ice Crystal Habit

As we indicated before, ice crystals of different habits may grow at different rates through either diffusional or collisional processes. In the present study, we also look into the effect of ice crystal habit on the cirrus development, assuming that the cloud consists of one of the following form types of ice crystals: hexagonal columns, hexagonal plates, bullet rosettes, or ice spheres. The first three are

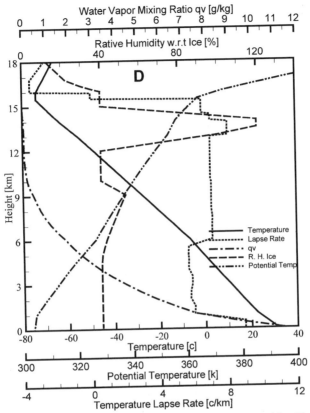

FIG. 6.9. Profiles of background atmospheric conditions for the cold-stable case.

commonly observed in cirrus clouds. The last kind has been used in many earlier studies for simplified ice microphysics modules and is used here as a reference case for comparison.

(4) Ventilation

In this model, we have implemented more accurate ventilation coefficients of nonspherical ice crystals as reported by Ji and Wang (1998). Because it would be useful to investigate the impact of ventilation, we have therefore performed and compared simulations with and without the ventilation effect.

For all simulations, we chose ammonium sulfate [$(NH_4)_2SO_4$] particles as the cloud condensation nuclei in our model. Although there is still debate about the

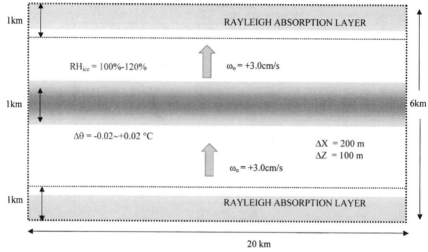

FIG. 6.10. Schematic of the model domain and initial configuration.

CCN types in cirrus, ammonium sulfate is currently thought to be the most important one.

6.4. Numerics of the Model

The model domain (Fig. 6.10) is two-dimensional and represents a cross section of cirrus advected by a nonsheared mean wind. The initial supersaturated layer is about 1 km thick. Temperatures in the supersaturated layer are randomly perturbed between $-0.02°C$ and $0.02°C$ relative to the base-state profile to generate an initial disturbance.

A weak large-scale updraft of 3 cm s^{-1} is imposed for all simulations. Vertical motions of this magnitude are very typical for fair-weather cirriform clouds. The temperature change resulting from the adiabatic expansion due to this small updraft is very weak and will not obscure the signal produced by the radiative effects. This makes it possible to examine the effect of radiation on the cirrus development.

The horizontal scale of the model is 20 km, and the vertical scale is 6 km. The spatial resolution used here is 200 m horizontally and 100 m vertically. In a test experiment in which the grid size was reduced to 100 m in the horizontal and 50 m in the vertical, the results did not show obvious differences from those obtained using the 200×100 m^2 grid cells.

Four time steps are applied to the different modules seperately: the radiative time step Δt_{rad}, the respective small and large dynamic time step $\Delta t_{dyn\text{-}small}$ and $\Delta t_{dyn\text{-}large}$, and the microphysical time step Δt_{mic}. In order to reduce the computational time spent on radiative transfer, the radiation time step is usually set larger than the dynamic time step (Lin, 1997). So the radiative transfer module is evaluated at 30 s, implying that the radiative heating rates remain constant during this 30 s.

Since our model system is quasi-compressible, the dynamic module is split into two parts, one for non–sound-wave related variables (large time step), the other is for sound-wave related variables (small time step). For computational stability, the dynamic time step has to meet the CFL (Courant–Friedrichs–Lewy) condition:

$$\Delta t < \frac{\Delta x}{V_{\max}} \tag{6.15}$$

where $V_{\max}$ is the maximum possible applicable phase speed including those of sound waves, falling particles, and the wind. Setting the dynamic step to 1 s requires that $V_{\max}$ be less than 100 m s^{-1}. This value is well above the terminal velocities of ice particles and the maximum wind speed in this study. As for the phase speed, the dispersion relation for linear internal gravity waves is

$$v = \pm \frac{Nk}{(k^2 + m^2)^{1/2}} \tag{6.16}$$

where v is frequency, k and m are the wave numbers in the horizontal and vertical directions respectively, and N is the Brunt–Väisälä frequency. The horizontal and vertical phase speeds are equal to v/k and v/m, respectively, with maximum possible phase speed about 3–4 m s^{-1}, also well below 100 m s^{-1}. Therefore, setting $\Delta t_{\text{dyn-large}}$ equal to 1 s is well suited to our situation. On the other hand, if we set $\Delta t_{\text{dyn-small}}$ equal to 0.1 s, $V_{\max}$ must be smaller than 900 m s^{-1}. This magnitude justifies the value of 0.1 s for the small dynamic time step in this study since the sound wave in our model has been reduced to 100 m s^{-1}.

In the microphysical module, the time step is set to 1.0 s. Theoretically, the time step for the diffusional growth of hydrometeors requires 0.2 s to ensure computational stability. We performed the test using 1.0 s, and the results did not show large differences from those using 0.2 s. Therefore, the time step for the microphysical module is set to 1.0 s.

6.5. Results and Discussion

6.5.1. Development of Cirrus Clouds in Different Atmospheric Environments

We first examine the development of cirrus clouds in the four different background environments. The ice crystals in these clouds are assumed to be columns for simplicity. Both solar (mainly shortwave) and terrestrial (mainly IR) radiations are included, implying that the results pertain to the daytime situation. In all cases, a background updraft of 3 m s^{-1} is imposed at all times. The microphysical and radiative profiles presented in the following sections are horizontally averaged values.

6.5.1.1. Warm-Unstable Case

Figure 6.11 shows the vertical profiles of ice water content (IWC) and the number concentration (N) of the cirrus cloud at 20, 40, 90, and 180 min. The top two panels show the total (pristine ice + aggregate) IWC and N, whereas the middle and lower two panels show these quantities for pristine ice and aggregates separately.

The top panels show that both IWC and N increase rapidly during the first 40 min, mainly due to homogeneous freezing nucleation. This nucleation process begins at 14.4 min and ends around 37 min. At each time the ice number concentration in general increases with height to a maximum somewhere in the upper cloud layer, then decreases sharply toward the cloud top. The peak number concentration of ice reaches its maximum value at about 40 min near the cloud top (9 km) before the decay starts. The decrease of ice concentration with height in the upper cloud layer is due to the fallout of ice crystals to lower layers and the termination of the nucleation process. The first mechanism is obviously triggered by the growth of ice crystals and their accompanying increases in fall speeds. The second mechanism is due to the diminishing water vapor supply, which is depleted by the rapid nucleation in the early stages of the cloud development.

The peak ice water content also is greatest at 40 min, as IWC values are generally greater in the upper parts of the cloud.

After 40 min, the peak values of IWC and N decrease while the thickness of the cloud increases from the original 1 km of the supersaturated layer depth to about 4 km at $t = 180$ min. This is undoubtedly due to the continuous settling of larger ice crystals, extending the cloud downward. While the ice concentration seems to redistribute more evenly in the vertical, the IWC has a pronounced relative minimum near 8 km at this time, essentially splitting the cloud into two distinct layers. This two-layer structure is most likely due to the increasing aggregation in the lower layer coupled with fairly active cloud development in the upper layer at the same time, as will be elaborated later.

The middle panels represent the time variation of the IWC and N profiles for pristine ice. It is clear that these curves mostly parallel those in the upper panels, reflecting the fact that pristine ice crystals are the dominant particles in the cloud, and the trends in the preceding two paragraphs apply to them as well.

The differences between the curves in the upper and middle panels, then, are caused by the behavior of the ice aggregates, as revealed in the lower panels. The lower left panel shows the behavior of the aggregate IWC. Aggregates do not form in the early stages because the ice crystals are small and the collisions between them are rare and inefficient. As ice crystals grow, however, collisions that result in aggregation become more frequent. Appreciable aggregation starts at about 40 min, modifying the IWC profile in the lower part of the cloud. The development of both N and IWC for aggregates become more vigorous and spreads in the vertical. Note that while the values of aggregate IWC are only a few times less

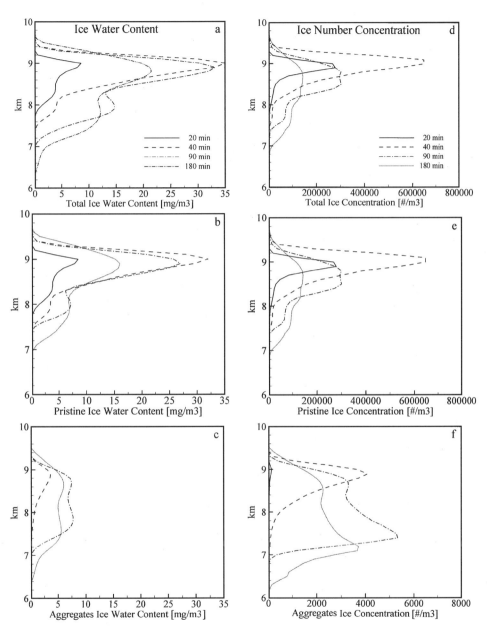

FIG. 6.11. Profiles of horizontal mean ice water content and horizontal mean ice number concentration for the warm-unstable case. Left panel: (a) Total ice water content, (b) pristine ice water content, (c) aggregate ice water content. Right panel: (d) Total ice number concentration, (e) pristine ice number concentration, (f) aggregate ice number concentration.

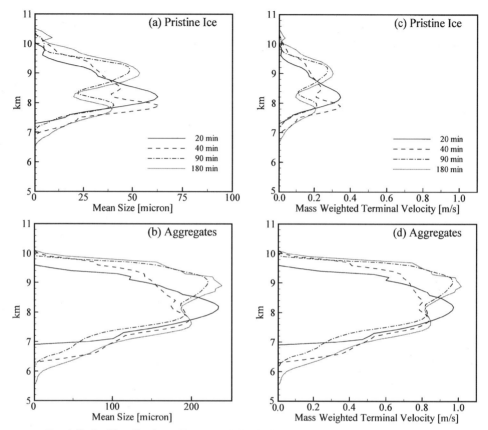

FIG. 6.12. Profiles of horizontally averaged distribution mean size and mass-weighted terminal velocity of ice crystals for the warm-unstable case: (a) Mean size of pristine ice, (b) mean size of aggregates, (c) terminal velocities of pristine ice, (d) terminal velocities of aggregates.

than those of pristine ice, the concentration values are about 100 times lower and hence do not significantly modify the total ice concentration profile. The aggregate concentration profile shows a more pronounced lower peak than the IWC profile.

Figure 6.12 shows the time variation of ice crystal size. Interestingly, the peak sizes of both pristine ice and aggregates do not seem to change appreciably over time, and only their positions are shifting. The pristine ice mean size profile shows a double-peak structure in the later stages of the cloud development, corresponding well to the total IWC profile. Even the aggregate size profile shows the double-peak structure, although the relative minimum is located slightly lower.

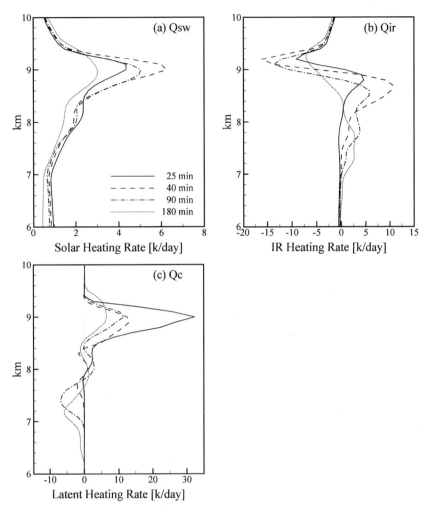

FIG. 6.13. Profiles of horizontally averaged (a) solar radiative heating rates (Q_{sw}), (b) IR hearing rates (Q_{ir}), and (c) latent heating rates (Q_c).

The effects of the cirrus cloud on radiation and latent heating are shown in Figure 6.13. Figure 6.13a shows the profile of the solar heating rate, which is sensitive to both the IWC and ice number concentration and becomes stronger in general when both those properties increase. Comparing Fig. 6.13a with the top two panels of Figure 6.11, we see that the solar heating rate profile generally parallels that of IWC and N, although the relation is nonlinear. Thus the strongest solar heating occurs at about 9 km.

The situation with the infrared heating is more complex. The cirrus cloud is "thin" in the sense of visible (shortwave) radiation, but not in the sense of IR. Ice particles in cirrus are efficient IR absorbers, and this is reflected in Figure 6.13b. Since IR enters the cloud from below, the lower parts of the cloud get heated first and the heating reaches a maximum at 8.8 km, lower down than the solar heating maximum. The peak heating rate, occurring at ~40 min, is ~11K/day, nearly twice that of the peak solar heating rate.

Instead of heating, the upper part of the cloud shows cooling in the IR band. Apparently, most terrestrial IR has been absorbed before reaching the upper cloud region, and the ice crystals there radiate more IR than they receive. It appears that, unlike for the solar radiation, which results in net heating, the IR heating and cooling in the cirrus cloud seem to cancel each other and thus result in little net heating or cooling. But it does have an important dynamic effect. The IR cooling maximum occurs at an altitude only slightly above the maximum solar heating, so that these two heating mechanisms tend to cancel each other at the same height level. However, the peak IR cooling rate, occurring at $t \sim 40$ min, is about 16K/day, more than compensating the peak solar heating of ~6K/day. Consequently, the net radiative effect of the upper cloud parts will be cooling, as will be elaborated later. For now, we note that peak warming in the lower parts and peak cooling in the upper parts of the cloud both occur at ~40 min. This warm-bottom/cool-top configuration obviously destabilizes the cloud layer and tends to promote stronger convection in the cloud if not compensated by other effects. But, as we have seen from the IWC and other microphysical profiles, this does not occur, implying that other factors do influence the cloud development.

Aside from microdynamic factors such as the aggregation and sedimentation of the ice crystals, as discussed previously, another factor of importance to the cloud development is latent heat. Figure 6.13c shows the time variation of the latent heating rate profile. Naturally, cooling occurs wherever there is evaporation.

In the early stages, the ice nucleation commences vigorously in the original supersaturated upper cloud layer and releases large amounts of latent heat in a short time. The latent heating rate at $t \sim 25$ min exceeds 30K/day, being nearly three times the IR heating rate and nearly five times the solar heating rate. Undoubtedly, the latent heating is the dominant factor.

6.5.1.2. Cold-Unstable Case

The results of the cold-unstable case look substantially different from those of the warm-unstable case. The time variation of the profiles for number concentration and water contents of total ice, pristine ice, and ice aggregates are plotted in Figure 6.14. Large amounts of ice crystals are produced in the upper cloud layers during the first 30 min due to rapid homogeneous freezing nucleation. The

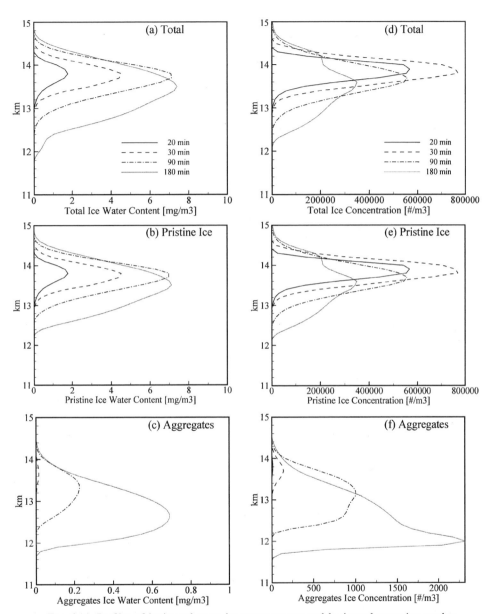

FIG. 6.14. Profiles of horizontal mean ice water content and horizontal mean ice number concentration for the cold-unstable case. Left panel: (a) Total ice water content, (b) pristine ice water content, (c) aggregate ice water content. Right panel: (d) Total ice number concentration, (e) pristine ice number concentration, (f) aggregate ice number concentration.

homogeneous nucleation begins at 13.7 min and ends around 33 min. It is activated again in the last 5 min of the simulation. The ice number concentration in general increases from the cloud base to a maximum value in the middle of the cloud, then decreases toward the cloud top, reaching its peak value near the cloud top (13.8 km) at about 30 min. Although it begins to decay after 30 min, the rate of decay is much slower than in the warm-unstable case. There is no double-layer structure, unlike in the warm-unstable case. In addition, the maximum concentration remains in the initial supersaturated layer to the end of the simulation. The behavior of the ice concentration may be the result of two processes. First, the heterogeneous nucleation process remains very active in the upper layer after the 30-min rapid growth period. This is because ice crystals grow more slowly at low temperatures and consume less water vapor per unit time, allowing more new ice crystals to nucleate through the heterogeneous nucleation process.The Second, the slow growth rates imply that the ice crystals remain small. maximum horizontally averaged mean sizes of pristine ice and aggregates are 20 μm and 100 μm, respectively (Fig. 6.15). Their fall velocities are small in this size range. Consequently, there are still many ice crystals remaining in the initial supersaturated layer by the end of the simulation. Although the ice number concentration starts to decay after 30 min, the ice water content keeps increasing long after, reaching its maximum value at about 150 min. This is because there are new ice crystals produced near the cloud top at all times and the ice crystal fallout rate is low due to their small sizes. Hence, most ice crystals continue to grow in the supersaturated layer.

Both the number concentration and IWC of aggregates at all levels are much smaller than those of pristine ice during the simulation, by the end of which the aggregates are concentrated in the lower parts of the cloud layer. The lowest 600 m of the cloud consists mostly of ice aggregates.

The profiles of the solar heating rates are shown in Figure 6.16a. Overall, the structure is similar to that of the warm-unstable case; namely, the maximum is near the cloud top. The peak solar heating occurs at around 30 min, again similar to the previous case.

The profiles of IR heating rates also change dramatically with time (Fig. 6.16b), but the behavior is significantly different from that of the warm-unstable case. The IR heating increases upward to a maximum value near 13.5 km level, then decreases toward the cloud top. The IR cooling at the cloud top is negligible, whereas the IR warming is significant throughout the whole cloud layer. This is because that the cirrus cloud simulated here is optically thin so that the IR radiation entering the cloud base can reach the cloud top. Thus, the whole cloud layer is radiatively heated during the simulation. The magnitude of IR heating rates is again much larger than solar heating. The maximum IR heating in the cloud can be as large as 40°C/day during the simulation. It is interesting to note that both IR and solar radiative heating rates seem to be more sensitive to ice number concentration than to IWC.

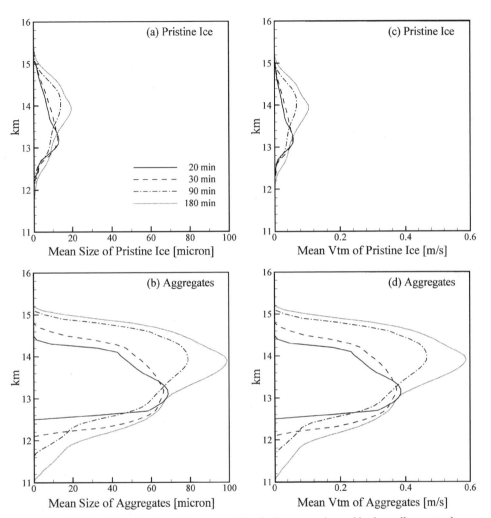

FIG. 6.15. Profiles of horizontally averaged distribution mean size and horizontally averaged mass-weighted terminal velocity of ice crystals for the cold-unstable case. (a) Mean size of pristine ice, (b) mean size of ice aggregates, (c) mass-weighted terminal velocity of pristine ice, (d) mass-weighted terminal velocity for ice aggregates.

The profiles of latent heating rates are shown in Figure 6.16c. The magnitude of latent heating is much smaller than that of the IR heating rate. This is caused by the small growth rates at cold temperatures that result in insignificant sublimation until 90 min into the simulation.

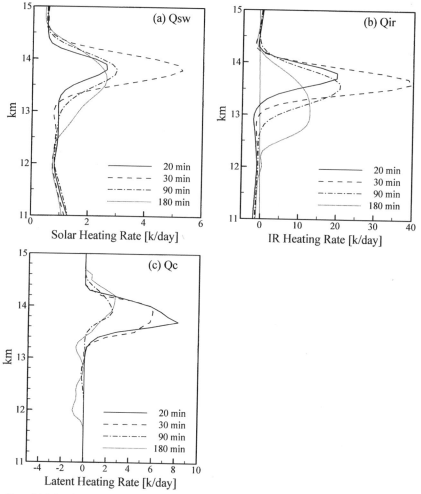

FIG. 6.16. Profiles of horizontally averaged (a) solar radiative heating rates (Q_{sw}), (b) IR hearing rates (Q_{ir}), and (c) latent heating rates (Q_c) for the cold-unstable case.

6.5.1.3. Warm-Stable Case

We now turn to the stable cases, examining the warm-stable case first. A special feature of the model results from this case, in contrast to the two previous cases: No homogeneous freezing nucleation takes place during the 180 min of simulation. This is mainly due to the initial stable structure that limits the vertical motion and hence the development of cloud and the vertical transport of water vapor available for the ice crystals to grow. Therefore, the cloud interior is not sufficiently cold or moist to favor homogeneous nucleation. As a result, the source of ice crystals for stable cases comes only from heterogeneous nucleation, which produces many fewer ice crystals than does the homogeneous nucleation process. The profiles of horizontally averaged ice water content and number concentration are shown in Figure 6.17. The location of maximum IWC moves rapidly downward with time due to the sedimentation during the first 40 min of the simulation. After that, the maximum stays fairly consistently at ~7.5 km.

The time variation of ice number concentration profile is more complicated. During the early simulation stages, the number concentration is distributed more or less uniformly within the initial supersaturated layer. This is because the initial relative humidity with respect to ice is uniform and only heterogeneous nucleation (which depends only on relative humidity) is involved. However, as time goes on, the number concentration changes dramatically. At later stages of the simulation, the ice crystals seem to concentrate in the subsaturated lower part of the cloud. This is because, at these stages, ice crystals grow larger when they fall down to the subsaturated environment so that most of them do not sublime to their core size. Consequently, the number of ice crystals does not decrease much due to sublimation in the lower part of the cloud. In fact, the number of ice crystals actually increases because ice crystals settling from the upper cloud layer are added to the ice concentration in the lower cloud layer. The increase of ice crystals due to fallout from the upper layer exceeds the number lost to sublimation.

Figure 6.18 shows the time variation of ice crystal sizes. It appears that, for both pristine ice and aggregates, the profiles remain fairly constant.

The profiles of solar, IR, and latent heating rates are shown in Figure 6.19. The magnitudes of these diabatic heating contributions are comparable. The IR heating rate (Fig. 6.19b) shows weak warming near the cloud base and cooling near the cloud top. The profiles of latent heating (Fig. 6.19c) indicate that the ice crystal fallout is significant after 40 min into the simulation. The total diabatic heating rates (Fig. 6.19d) show warming near the cloud base and cooling near the cloud top after 40 min. In this simulation, the diabatic heatings do not stabilize the cloud layer. This is consistent with the profiles of equivalent potential temperature (not shown). In this case, the effect of diabatic heating does not exceed that of adiabatic cooling.

6.5.1.4. Cold-Stable Case

Just as in the warm-stable case, homogeneous nucleation does not occur in this case. The profiles of IWC and ice number concentration are similar to those for the warm stable case, but their time variations are different (Fig. 6.20). The major

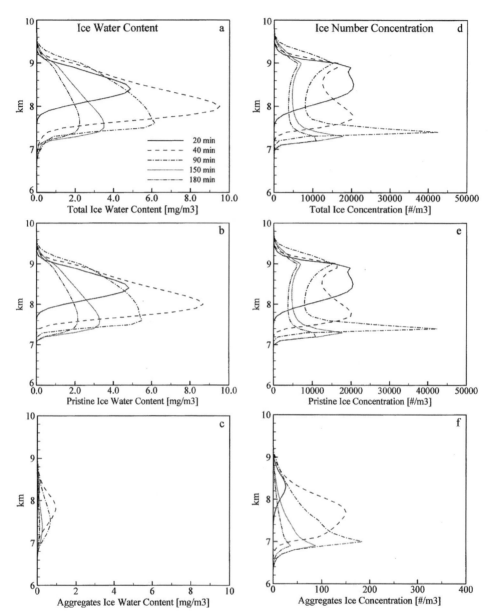

FIG. 6.17. Profiles of horizontal mean ice water content and horizontal mean ice number concentration for the warm-stable case. Left panel: (a) Total ice water content, (b) pristine ice water content, (c) aggregate ice water content. Right panel: (d) Total ice number concentration, (e) pristine ice number concentration, (f) aggregate ice number concentration.

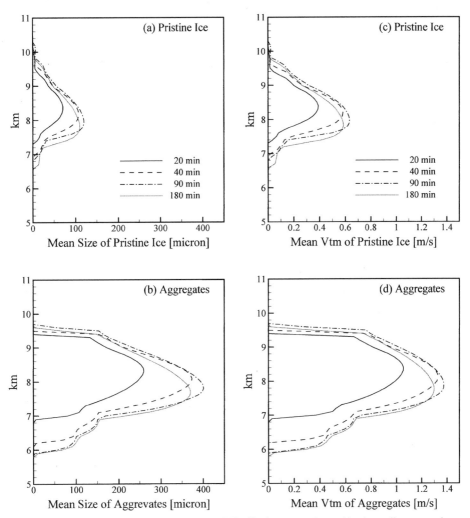

FIG. 6.18. Profiles of horizontally averaged distribution mean size and horizontally averaged mass-weighted terminal velocity of ice crystal for the warm-stable case. (a) Mean size of pristine ice, (b) mean size of ice aggregates, (c) mass-weighted terminal velocity of pristine ice, (d) mass-weighted terminal velocity for ice aggregates.

cause of the differences is the sensitivity of the growth rates to temperature. The growth rates of ice crystals at cold temperatures are very small so that ice crystals remain small and stay in the initial supersaturated layer longer. Not until they grow to sizes with substantial terminal velocities would they fall out of this layer.

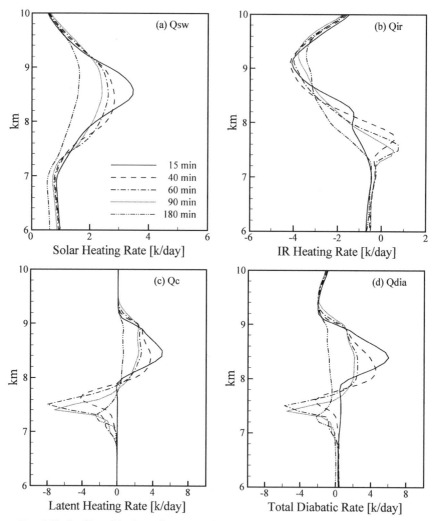

FIG. 6.19. Profiles of horizontally averaged (a) solar radiative heating rates (Q_{sw}), (b) IR heating rates (Q_{ir}), (c) latent heating rates (Q_c), and (d) total diabatic heating rates (Q_{dia}) for the warm-stable case.

Hence, it takes substantially longer for the ice crystals to grow in this case and the maximum IWC does not peak until 90 min.

Figure 6.21 shows the mean size profiles for pristine ice and aggregates. The mean sizes in both categories are only about half the corresponding values for the warm-stable case, confirming the reasoning in the above paragraph.

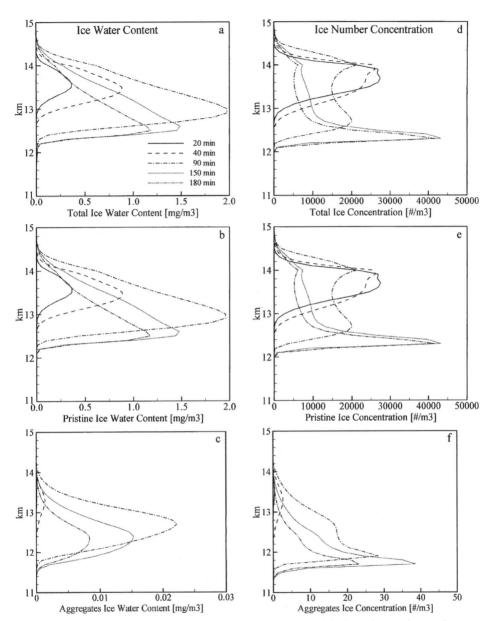

FIG. 6.20. Profiles of horizontal mean ice water content and horizontal mean ice number concentration for the cold-stable case. Left panel: (a) Total ice water content, (b) pristine ice water content, (c) aggregate ice water content. Right panel: (d) Total ice number concentration, (e) pristine ice number concentration, (f) aggregate ice number concentration.

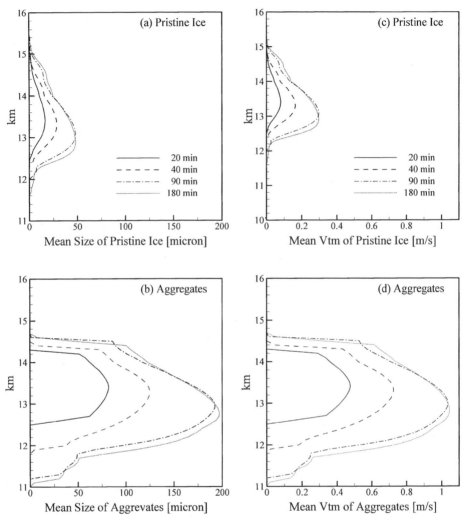

FIG. 6.21. Profiles of horizontally averaged distribution mean size and horizontally averaged mass-weighted terminal velocity of ice crystals for the cold-stable case. (a) Mean size of pristine ice, (b) mean size of ice aggregates, (c) mass-weighted terminal velocity of pristine ice, (d) mass-weighted terminal velocity for ice aggregates.

The profiles of solar heating, IR heating, and latent heating rates are shown in Figure 6.22. Both solar and IR heating rates are very sensitive to the IWC, and less so to ice number concentration. They are largest around 70 min into simulation. The profiles of IR heating show IR warming at the cloud base.

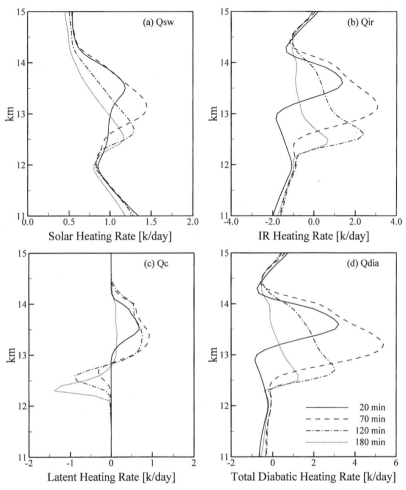

FIG. 6.22. Profiles of horizontally averaged (a) solar radiative heating rates (Q_{sw}), (b) IR hearing rates (Q_{ir}), (c) latent heating rates (Q_c), and (d) total diabatic heating rates (Q_{dia}) for the cold-stable case.

The IR cooling is distributed throughout most of the cloud deck due to small optical depth. The strength of the warming effect is decreasing upward, and eventually becomes weak cooling near the cloud top. The profiles of latent heating reveal that the ice crystal fallout is not significant until 70 min. The IR heating within the cloud layer dominates the total diabatic heating rates. Solar heating and IR warming balance the cooling due to sublimation in the lower part of the cloud. The diabatic heating does not destabilize the cloud layer during the cloud development.

6.5.2. Effect of Ice Crystal Habit on the Development of Cirrus Clouds

Do ice crystal habits in a cirrus cloud influence its development? This is the problem to be examined in this section. Four types of ice crystals are considered in the simulation set presented herein: columns, plates, bullet rosettes, and spheres. The first three types are commonly observed in cirrus clouds, while the last one is the simplified shape as an approximation to ice particles in many previous models. To focus on the effect of habit, we will first consider the development of cirrus in cold-stable and warm-stable atmospheres, but turn off the aggregation process. Then we perform the simulation on warm-unstable and cold-unstable atmospheres with the aggregation process turned on. A summary of the scenarios simulated is given in Table 6.8.

The sensitivity of cirrus development to ice crystal habit in a stable environment can be seen in Figures 6.23 and 6.24 for warm and cold cirrus, respectively. For the warm-stable cases (Fig. 6.23), the IWCs reach their peak values around 40 min. However, the magnitudes of the peak values for different habits are quite different. The peak IWC for cirrus consisting of rosettes is more than twice as large as the one for spheres. The peak values of ice water content for columns and plates are similar. They are greater than for spheres, but smaller than for rosettes. The results for the cold-stable case (Fig. 6.24) are similar, except that the IWC reaches its maximum value at around 90 min instead of 40 min into the simulation.

TABLE 6.8 SUMMARY OF SCENARIOS SIMULATED TO TEST THE EFFECT OF CRYSTAL HABIT

Index	Ice crystal habit	Temperature zone	Static stability	Aggregation process	Radiative process	Vertical velocity (cm/s)
1	Columns	Warm	Stable	No	SW+IR	3
2	Plate	Warm	Stable	No	SW+IR	3
3	Rosettes	Warm	Stable	No	SW+IR	3
4	Spheres	Warm	Stable	No	SW+IR	3
5	Columns	Cold	Stable	No	SW+IR	3
6	Plate	Cold	Stable	No	SW+IR	3
7	Rosettes	Cold	Stable	No	SW+IR	3
8	Spheres	Cold	Stable	No	SW+IR	3
9	Columns	Warm	Unstable	Yes	SW+IR	3
10	Rosettes	Warm	Unstable	Yes	SW+IR	3
11	Columns	Cold	Unstable	Yes	SW+IR	3
12	Rosettes	Cold	Unstable	Yes	SW+IR	3

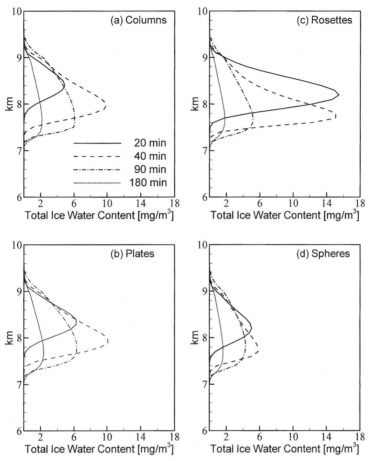

FIG. 6.23. Warm-stable case. Profiles of horizontally averaged ice water content for four different ice crystal types: (a) Columns, (b) plates, (c) bullet rosettes, (d) spheres.

The corresponding profiles of mean size for the warm-stable and cold-stable cases are shown in Figures 6.25 and 6.26, respectively. The largest mean size of ice crystals is for rosettes, followed by plates and columns. The mean size of spheres is the smallest. The difference in mean size can be as large as fourfold between the cases of rosettes and spheres.

The corresponding radiative heating rates for the warm-stable and cold-stable cases are shown in Figures 6.27 through 6.30. In all cases, rosettes have the biggest impact. This is especially true for IR heating, whose peak rate for rosettes (~15K/day) can be more than 10 times that for spheres. This indicates that, at least

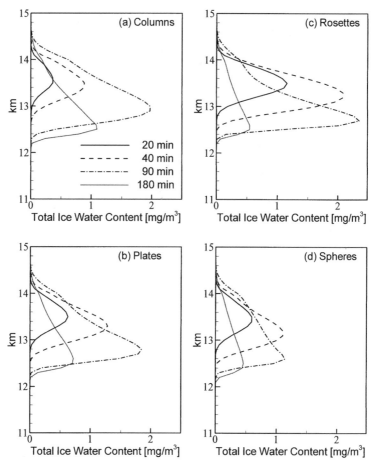

FIG. 6.24. Cold-stable case. Profiles of horizontally averaged ice water content for four different ice crystal types: (a) Columns, (b) plates, (c) bullet rosettes, (d) spheres.

as far as radiation is concerned, assuming spherical ice particles in cirrus clouds will lead to large errors. Although the resulting profiles of IWC for columns and plates are similar, the corresponding solar and IR heating rates are somewhat larger for plates than for columns. It can be seen that the IR heating rate is more sensitive to ice crystal habit than is the solar radiative heating rate.

The differences among simulations with different ice habit may be explained by ice crystal capacitance. As explained in Section 6.2.2, the capacitance is parameterized as a function of the maximum dimension. A plot of the capacitance as a function of mass and habit is shown in Figure 6.31. For ice particles with

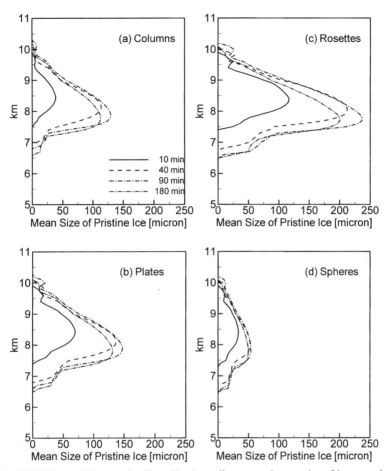

FIG. 6.25. Warm-stable case. Profiles of horizontally averaged mean size of ice crystals for four different ice crystal types: (a) Columns, (b) plates, (c) bullet rosettes, (d) Spheres.

the same mass, bullet rosettes have the largest capacitance, and hence the greatest diffusional growth rates, followed by plates and columns, whereas spheres have the smallest capacitance. Results here show that the IWC, in general, varies similarly with ice crystal habit. Although the capacitance of columns is larger than that of plates, the resulting IWC is almost the same (Figs. 6.23–6.24). This may be because the differential radiative heating induced by plates is greater than that by columns, resulting in slightly stronger upward vertical motion for plates and thus producing more IWC than expected.

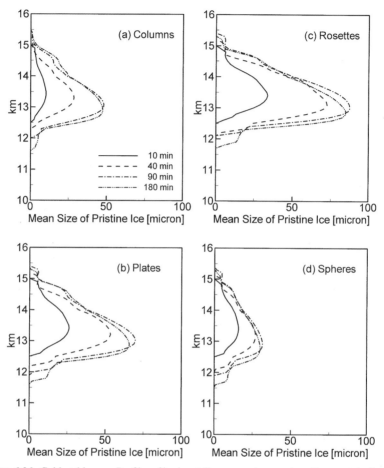

FIG. 6.26. Cold-stable case. Profiles of horizontally averaged mean size of ice crystals for four different ice crystal types: (a) Columns, (b) plates, (c) bullet rosettes, (d) spheres.

6.5.3. Effect of Ventilation on the Development of Cirrus Clouds

We have seen in Section 6.2.2 that the ventilation coefficient is a sensitive function of ice crystal habit. Among the four ice crystal habits considered here, the column has the greatest ventilation coefficient for the same X parameter. In this study, we chose columns to show how sensitive the development of cirrus is to the ventilation effect. The scenarios are summarized in Table 6.9.

The profiles of IWC for simulations with and without the ventilation effect are shown in Figure 6.32. The ventilation effect is more evident during the later period of the simulation, and also for the warm-unstable case than for the cold-unstable

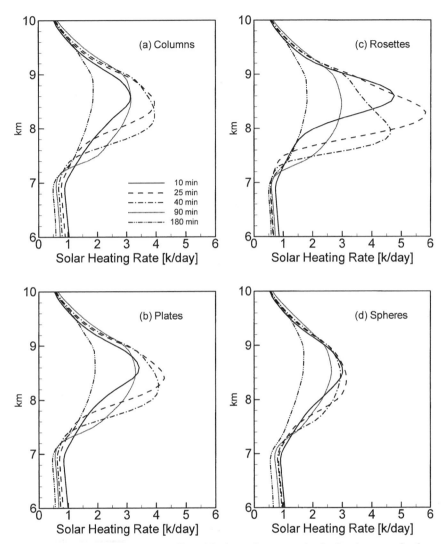

FIG. 6.27. Warm-stable case. Profiles of horizontally averaged solar heating rates for four different ice crystal types: (a) Columns, (b) plates, (c) bullet rosettes, (d) spheres.

case. It is known from Section 6.5.1 that the cold cirrus clouds are characterized by smaller ice crystals than warm cirrus clouds. Since larger ice crystals usually have greater terminal velocities, they would show a more pronounced ventilation effect. The difference can be as large as 25% for the IWC by the end of the simulation for the warm-unstable case. The domain-averaged IWC for the warm and cold

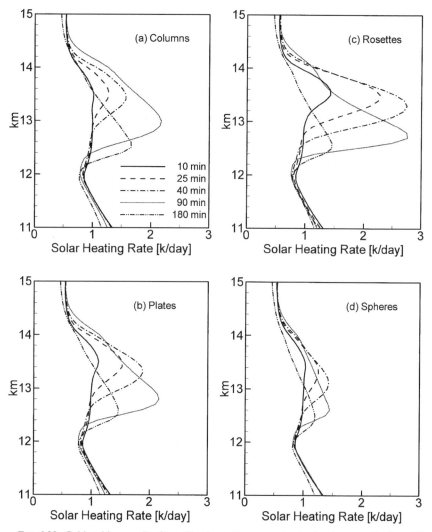

FIG. 6.28. Cold-stable case. Profiles of horizontally averaged solar heating rates for four different ice crystal types: (a) Columns, (b) plates, (c) bullet rosettes, (d) spheres.

cases are shown in Figure 6.33. Since the cirrus clouds considered in this study are thin and consist of smaller ice crystals than in more convective cases (e.g., those associated with convective clouds), the ventilation effect is apparently not very strong. It is seen, however, that the effect of ventilation cannot be ignored, and is most pronounced during the later stages of the simulation.

TABLE 6.9 SUMMARY OF SCENARIOS SIMULATED TO TEST THE VENTILATION EFFECT

Index	Ice crystal habit	Temperature zone	Static stability	Aggregation process	Diabatic process	Vertical velocity (cm/s)
1	Columns	Warm	Unstable	Yes	SW+IR Daytime	3
2	Columns	Warm	Unstable	Yes	Daytime No Q_c	3
3	Columns	Warm	Unstable	Yes	No	3

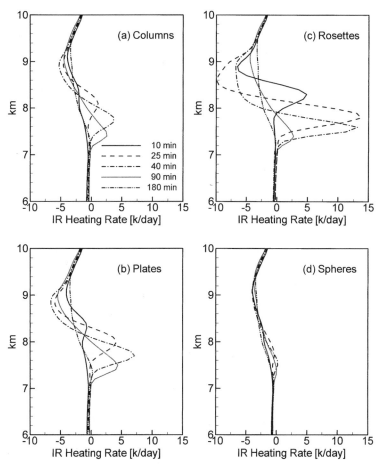

FIG. 6.29. Warm-stable case. Profiles of horizontally averaged IR heating rates for four different ice crystal types: (a) Columns, (b) plates, (c) bullet rosettes, (d) spheres.

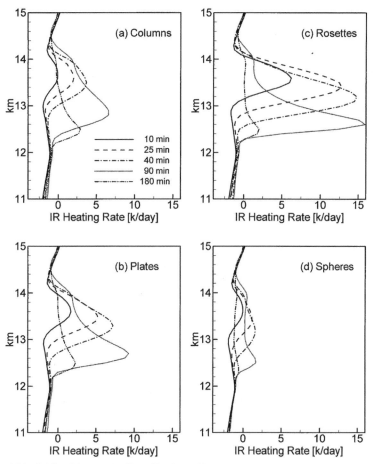

FIG. 6.30. Cold-stable case. Profiles of horizontally averaged IR heating rates for four different ice crystal types: (a) Columns, (b) plates, (c) bullet rosettes, (d) spheres.

Another reason why the ventilation effect is not very pronounced in the cases studied here is that all the cirrus clouds examined here developed in relatively quiescent environments. Remember that the background perturbation vertical velocity is only 3 cm s^{-1}, whereas in the vicinity of a deep convective storm that can produce extensive cirrus anvil clouds, the background perturbation vertical velocity can easily exceed this value by tenfold or more, in which case the ice crystals will grow to much larger sizes and the ventilation effect is expected to play an important role in the development of the clouds.

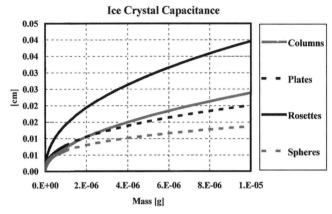

FIG. 6.31. Ice crystal capacitance as a function of ice crystal mass for different habits.

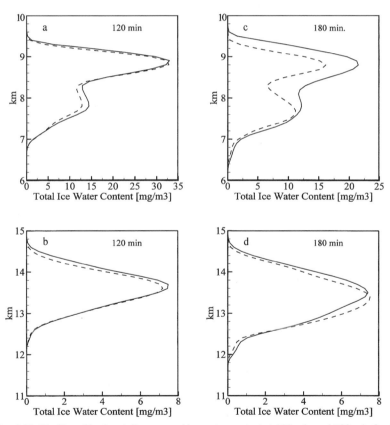

FIG. 6.32. Profiles of horizontally averaged ice water content at 120 min and 180 min for the warm-unstable case (top) and the cold-unstable case (bottom). The solid line is for simulation with ventilation effect; the dashed line is for simulation without ventilation effect.

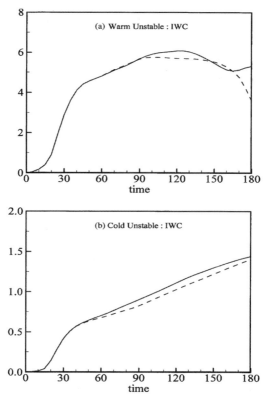

FIG. 6.33. Domain-averaged ice water content for (a) the warm-unstable case and (b) the cold-unstable case. The solid line is for simulation with ventilation effect; the dashed line is for simulation without ventilation effect.

APPENDIX A

AREA OF AN AXIAL CROSS SECTION

Equation (2.39) gives

$$A = 2 \int_{-c}^{c} x \, dz = 2a \int_{-c}^{c} \sqrt{1 - \frac{z^2}{C^2}} \, \cos^{-1}\left(\frac{z}{\lambda C}\right) dz$$

If we let $u = z/C$, then

$$A = 2aC \int_{-1}^{1} \sqrt{1 - u^2} \, \cos^{-1}\left(\frac{u}{\lambda}\right) du \tag{A.1}$$

Next, let

$$I = \int_{-1}^{1} \sqrt{1 - u^2} \, \cos^{-1}\left(\frac{u}{\lambda}\right) du \tag{A.2}$$

Since $|u| < 1$, $|u|/\lambda < 1$, we can expand the arccosine function into an infinite series in (u/λ):

$$\cos^{-1}\left(\frac{u}{\lambda}\right) = \left(\frac{\pi}{2}\right) - \left[\sum_{n=0}^{\infty} \frac{(2n-1)!}{(2n)!\,(2n+1)} \left(\frac{u}{\lambda}\right)^{2n+1}\right]$$

$$= \frac{\pi}{2} - \left[\left(\frac{u}{\lambda}\right) + \frac{1}{2\cdot 3}\left(\frac{u}{\lambda}\right)^3 + \frac{1\cdot 3}{2\cdot 4\cdot 5}\left(\frac{u}{\lambda}\right)^5\right.$$

$$\left. + \frac{1\cdot 3\cdot 5}{2\cdot 4\cdot 6\cdot 7}\left(\frac{u}{\lambda}\right)^7 + \cdots\right]$$

$$= \frac{\pi}{2} - \left[\left(\frac{1}{\lambda}\right)u + \left(\frac{0.167}{\lambda^3}\right)u^3 + \left(\frac{0.0752}{\lambda^5}\right)u^5\right.$$

$$\left. + \left(\frac{0.0446}{\lambda^7}\right)u^7 + \cdots\right] \tag{A.3}$$

Putting (A.3) into (A.2), we obtain

$$I = \frac{\pi}{2}\int_{-1}^{1}\sqrt{1 - u^2}\,du - \left(\frac{1}{\lambda}\right)\int_{-1}^{1} u\sqrt{1 - u^2}\,du$$

$$- \left(\frac{0.167}{\lambda^3}\right)\int_{-1}^{1} u^3\sqrt{1 - u^2}\,du + \cdots \tag{A.4}$$

But for $m = 2n + 1$ $(n = 0, 1, 2, \ldots)$,

$$\int_{-1}^{1} u^m\sqrt{1 - u^2}\,du = -\frac{1}{(m+2)}\left[u^{m-1}\left(\sqrt{1 - u^2}\right)^3\right]_{-1}^{1}$$

$$- (m-1)\int_{-1}^{1} u^{m-2}\sqrt{1 - u^2}\,du \tag{A.5}$$

(see, for example, Dwight, 1961). Therefore

$$I = \frac{\pi}{2}\int_{-1}^{1}\sqrt{1 - u^2}\,du = \frac{\pi}{2}\left[\frac{u\sqrt{1 - u^2}}{2} + \frac{1}{2}\,\sin^{-1}u\right]_{-1}^{1} = \frac{\pi^2}{4}$$

and

$$A = 2aCI = \frac{\pi^2}{2}aC \tag{A.6}$$

Equation (A.6) can be easily checked for two special cases: $\lambda = 1$ and $\lambda = \infty$.

(i) $\lambda = 1$. In this case we let $v = \cos^{-1} u$. Then, from (A.2),

$$I = \int_{-1}^{1} \sqrt{1 - u^2} \, \cos^{-1} u \, du = - \int_{\pi}^{0} v \, \sin^2 v \, dv$$

$$= - \left[\frac{v^2}{4} - \frac{v \sin 2v}{4} - \frac{\cos 2v}{8} \right]_{\pi}^{0} = \frac{\pi^2}{4}$$

Thus,

$$A_{\lambda=1} = \frac{\pi^2}{2} aC \tag{A.7}$$

(ii) $\lambda = \infty$. In this case $\cos^{-1}(u/\lambda) = \cos^{-1}(0) = \pi/2$ and

$$I = \frac{\pi}{2} \int_{-1}^{1} \sqrt{1 - u^2} \, du = - \frac{\pi}{2} \int_{\pi}^{0} \sin^2 v = - \frac{\pi}{2} \left[\frac{v}{2} - \frac{\sin^2 v}{4} \right]_{\pi}^{0} = \frac{\pi}{4}$$

Thus

$$A_{\lambda=\infty} = \frac{\pi^2}{2} aC \tag{A.8}$$

APPENDIX B

CALCULATION OF VOLUME

Equation (2.41) gives

$$V = \pi a^2 \int_{-C}^{C} \left(1 - \frac{z^2}{C^2} \right) \left[\cos^{-1} \left(\frac{z}{\lambda C} \right) \right]^2 dz$$

$$= \pi a^2 C \int_{-1}^{1} (1 - u^2) \left[\cos^{-1} \left(\frac{u}{\lambda} \right) \right]^2 du \tag{B.1}$$

where $u = z/C$. We investigate three cases of λ:

(i) $\lambda = 1$. Equation (B.1) becomes (letting $v = \cos^{-1} u$)

$$V = \pi a^2 C \int_{-1}^{1} (1 - u^2) \, [\cos^{-1} u]^2 \, du$$

$$= -\pi a^2 C \int_{\pi}^{0} v^2 \sin^3 v \, dv = \pi a^2 C \left[\frac{1}{27} - \frac{\pi^2}{12} - 3 + \frac{3}{4}\pi^2 \right]$$

$$= 3.6167 \pi a^2 C$$

(ii) $\lambda = \infty$. Equation (B.1) becomes

$$V = \pi a^2 C \left(\frac{\pi}{2}\right)^2 \int_{-1}^{1} (1 - u^2) \, du = \frac{\pi^3}{4} a^2 C \left(2 - \frac{2}{3}\right) = \frac{\pi^2}{3} \pi a^2 C$$

$$= 3.2889 \pi a^2 C$$

(iii) $1 < \lambda < \infty$. In this case we expand $\cos^{-1}(u/\lambda)$ as in Eq. (A.3). Squaring it, we have

$$\left[\cos^{-1}\left(\frac{u}{\lambda}\right)\right]^2 = \frac{\pi^2}{4} - \left(\frac{\pi}{\lambda}\right)u + \left(\frac{1}{\lambda^2}\right)u^2 - \left(\frac{0.167\pi}{\lambda^3}\right)u^3$$

$$+ \left(\frac{0.333}{\lambda^4}\right)u^4 - \left(\frac{0.075\pi}{\lambda^5}\right)u^5 + \left(\frac{0.1783}{\lambda^6}\right)u^6 - \cdots$$

$$\text{(B.2)}$$

Equation (B.1) then becomes

$$V = \pi a^2 C \int_{-1}^{1} (1 - u^2) \left[\frac{\pi^2}{4} - \left(\frac{\pi}{\lambda}\right)u + \left(\frac{1}{\lambda^2}\right)u^2 - \left(\frac{0.167\pi}{\lambda^3}\right)u^3\right.$$

$$\left. + \left(\frac{0.333}{\lambda^4}\right)u^4 - \cdots\right] du$$

Integration of terms with odd powers of u in the square brackets yields even functions of u, therefore, the odd terms vanish. Only terms with even powers of u survive. The result is

$$V = \pi a^2 C \left[3.2889 + \frac{0.2667}{\lambda^2} + \frac{0.0381}{\lambda^4} + \frac{0.0113}{\lambda^6} + \frac{0.0046}{\lambda^8} + \frac{0.0024}{\lambda^{10}} + \cdots\right]$$

$$\text{(B.3)}$$

If we neglect higher-order terms, we see that Eq. (B.3) becomes identical with Eq. (2.44).

APPENDIX C

CLOSED-FORM EXPRESSION OF THE CONICAL VOLUME

A conical particle whose axial cross section is described by Eq. (2.25) has a volume (assuming rotational symmetry with respect to the z-axis) given by

$$V = \pi a^2 \int_{-c}^{c} \left(1 - \frac{z^2}{c^2}\right) \left[\cos^{-1}\left(\frac{z}{\lambda c}\right)\right]^2 dz \qquad \text{(C.1)}$$

In Appendix B, this volume integral is approximated by a series of expansions.

It turns out that the integral (C.1) can be calculated in the quadratures to yield a closed-form expression as given below.

By defining

$$\xi = \frac{z}{c} \tag{C.2}$$

Equation (C.1) can be rewritten as

$$V = \pi a^2 c \int_{-1}^{1} (1 - \xi^2) \left[\cos^{-1} \left(\frac{z}{\lambda c} \right) \right]^2 d\xi \tag{C.3}$$

Consider the integral

$$I = \int_{-1}^{1} (1 - \xi^2) \left[\cos^{-1} \left(\frac{z}{\lambda c} \right) \right]^2 d\xi = \int_{1}^{1} (1 - \xi^2) \left\{ \left[\cos^{-1} \left(\frac{z}{\lambda c} \right) \right]^2 \right.$$
$$\left. + \left[\cos^{-1} \left(\frac{z}{\lambda c} \right) \right]^2 \right\} d\xi \tag{C.4}$$

The quantity in the wavy brackets in the second integral can be reduced further to become

$$\frac{\pi^2}{2} + 2 \left(\sin^{-1} \left(\frac{\xi}{\lambda} \right) \right)^2 \tag{C.5}$$

so that Eq. (C.4) becomes

$$I = \left(\frac{\pi^2}{3} \right) + 2 \int_{0}^{1} (1 - \xi^2) \left(\sin^{-1} \left(\frac{\xi}{\lambda} \right) \right)^2 d\xi \tag{C.6}$$

Using the technique of integration in parts with

$$u = \left(\sin^{-1} \left(\frac{\xi}{\lambda} \right) \right)^2$$

and

$$dv = (1 - \xi^2) \, d\xi$$

yields

$$I_1 = \int_{0}^{1} (1 - \xi^2) \left(\sin^{-1} \left(\frac{\xi}{\lambda} \right) \right)^2 d\xi = \frac{2}{3} \left(\sin^{-1} \left(\frac{\xi}{\lambda} \right) \right)^2 \Big|_{0}^{1}$$
$$- 2 \int_{0}^{1} \frac{\xi (\sin^{-1}(\xi/\lambda))}{\sqrt{\lambda^2 - \xi^2}} \, d\xi + \frac{2}{3} \int_{0}^{1} \frac{\xi^3 (\sin^{-1}(\xi/\lambda))}{\sqrt{\lambda^2 - \xi^2}} \, d\xi \tag{C.7}$$

where, by using the same technique, we can get

$$I_2 = \int_0^1 \frac{\xi(\sin^{-1}(\xi/\lambda))}{\sqrt{\lambda^2 - \xi^2}} d\xi = -\sqrt{\lambda^2 - 1} \; \sin^{-1}\left(\frac{1}{\lambda}\right) + 1 \qquad (C.8)$$

and

$$I_3 = \int_0^1 \frac{\xi^3(\sin^{-1}(\xi/\lambda))}{\sqrt{\lambda^2 - \xi^2}} d\xi = -\left(\frac{2\lambda^2 + 1}{3}\right)\sqrt{\lambda^2 - 1} \; \sin^{-1}\left(\frac{1}{\lambda}\right)$$
$$+ \left(\frac{1}{3}\right)\left(2\lambda^2 + \frac{1}{3}\right) \qquad (C.9)$$

Thus the volume integral finally becomes

$$V = \pi a^2 c \left\{ \frac{\pi^2}{3} + \frac{4}{3}\left(\sin^{-1}\left(\frac{1}{\lambda}\right)\right)^2 - \left(\frac{8\lambda^2 - 32}{9}\right)\sqrt{\lambda^2 - 1} \; \sin^{-1}\left(\frac{1}{\lambda}\right) \right.$$
$$\left. + \left(\frac{24\lambda^2 - 104}{27}\right) \right\} \qquad (C.10)$$

It is useful to check if this expression converges to proper limits for the two extreme values of λ. For $\lambda = 1$, we have

$$V = \pi a^2 c \left(\frac{2\pi^2}{3} - \frac{80}{27}\right) = 3.6167\pi a^2 c \qquad (C.11)$$

For $\lambda \to \infty$, the sum of all terms in the curly brackets other than the first in Eq. (C.9) vanishes, as can be shown by taking its limit as $\lambda \to \infty$. Hence

$$V = \pi a^2 c \left(\frac{\pi^2}{3}\right) \qquad (C.12)$$

Equations (C.11) and (C.12) are identical to those given by Wang (1982; Appendix B) via normal derivation. The closed-form expression (C.10) thus converges to proper forms at these two extremes.

References

Ackerman, S. A., and Stephens, G. L. (1987). The absorption of solar radiation by cloud droplets: An application of anomalous diffraction theory. *J. Atmos. Sci.* **44**, 1574–1588.

Ackerman, T. P., Liou, N. K., Valero, F. P. J., and Pfister, L. (1988). Heating rates in tropical anvils. *J. Atmos. Sci.* **45**, 1606–1623.

Anderson, D. A., Tannehill, J. C., and Pletcher, R. H. (1984). "Computational Fluid Mechanics and Heat Transfer." McGraw-Hill, New York.

Auer, H. A., and Veal, D. L. (1970). The dimensions of ice crystals in natural clouds. *J. Atmos. Sci.* **27**, 919–926.

Battan, L. (1973). "Radar Observation of the Atmosphere." University of Chicago Press.

Bauer, C. F., and Pitter, R. L. (1982). Investigation of riming and snowflake electrification in strati-form clouds. *Precip. Cloud Physics Conf.*, Chicago, pp. 568–569.

Beard, K. V., and Pruppacher, H. R. (1971). A wind tunnel investigation of the rate of evaporation of small water drops falling at terminal velocity in air. *J. Atmos. Sci.* **28**, 1455–1464.

Bentley, W. A., and Humphreys, W. J. (1931). "Snow Crystals." Dover, New York.

Bentley, W. A., and Humphreys, W. J. (1962). Snow Crystals. Dover, 227pp.

Böhm, J. P. (1989). A general equation for the terminal fall speed of solid hydrometeors. *J. Atmos. Sci.* **46**, 2419–2427.

Braza, M., Chassiang, P., and Minh, H. Ha (1986). Numerical study and physical analysis of the pressure and velocity fields in the near wake of a circular cylinder. *J. Fluid Mech.* **165**, 79–130.

Brenner, H. (1963). Forced convection heat and mass transfer at small Peclect numbers from a particle of arbitrary shape. *Chem. Eng. Sci.* **18**, 109–122.

Bruntjes, R. T., Heymsfield, A. J., and Krauss, T. W. (1987). An examination of double-plate ice crystals and the initiation of precipitation in continental cumulus clouds. *J. Atmos. Sci.* **44**, 1331–1349.

Carnuth, W. (1967). Zur Abhabengigheit des Aerosol Partikel Spektrum von Meteorologischen Vorgaengen und Zustaende. *Arch. Meteor. Geophsik. Biok. A*, **16**, 321–343.

Carrier, G. (1953). On slow viscous flow. Final Rept., Office of Naval Research Contract Nonr-653(00), Brown University, 31 pp. (A.D. 16588).

Cess *et al.* (1989). Interpretation of cloud-climate feedback as produced by 14 atmospheric general circulation models. *Science* **245**, 513–515.

Chagnon, C. W., and Junge, C. E. (1961). The vertical distribution of aerosol particles in the stratosphere. *J. Meteorol.* **18**, 746–752.

Chilukuri, R. (1987). Incompressible laminar flow past a transversely vibrating cylinder. *Trans. ASME* **109**, 166–171.

Clift, R., Grace, J. R., and Weber, M. E. (1978). "Bubbles, Drops, and Particles." Academic Press, New York.

Cotton, W. R., and Anthes, R. A. (1989). "Storm and Cloud Dynamics." Academic Press.

Courant, R., and John, F. (1965). "Introduction to Calculus and Analysis," Vol. I. Wiley-InterScience, New York.

Cox, S. K. (1971). Cirrus clouds and the climate. *J. Atmos. Sci.* **28**, 1513–1515.

Dalle Valle, J. M., Orr, C., and Hinkle, B. L. (1954). A new method for the measurement of aerosol electrification. *Brit. J. Appl. Phys. Suppl.* **3**, S198–S206.

Davis, R. W. (1984). Finite difference methods for fluid flow. Computational Techniques and Applications, J. Noye and C. Fletcher, Eds., Elsevier, 51–59.

DeMott, P. J., Meyers, M. P., and Cotton, W. R. (1994). Numerical model simulations of cirrus clouds including homogeneous and heterogeneous ice nucleation. *J. Atmos. Sci.* **51**, 77–90.

Dennis, S. C. R., and Chang, G. Z. (1969). Numerical integration of the Navier-Stokes equations for two-dimensional flow. *Phys. Fluid Suppl.* **II**, 88–93.

Dennis, S. C. R., and Chang, G. Z. (1970). Numerical solutions for steady flow past a cylinder at Reynolds numbers up to 1000. *J. Fluid Mech.* **42**, 471–489.

Deshler, T. L. (1982). Contact nucleation by submicron atmospheric aerosols. Preprints Conf. Cloud Physics, Chicago, pp. 111–114.

D'Errico, R. E., and Auer, A. H. (1978). An observational study of the accretional properties of ice crystals of simple geometric shapes. In Preprints Conf. Cloud Physics and Atmos. Electr., Issaquah, WA, 114–121.

Devulapalli, S. S. N., and Collett Jr., J. L. (1994). The influence of riming and frontal dynamics on winter precipitation chemistry in level terrain. *Atmos. Res.* **32**, 203–213.

Dwight, H. B. (1961). "Tables of Integrals and Other Mathematical Data," 4th Ed. Macmillan.

Einstein, A. (1905). *Ann. d. Physik.* **17**, 549.

Flatau, P. J., Walko, R. L., and Cotton, W. R. (1992). Polynomial fits to saturation vapor pressure. *J. Appl. Meteorol.* **31**, 1507–1513.

Freitas, C. J., Street, R. L., Findikakis, A. N., and Koseff, J. R. (1985). Numerical simulation of three-dimensional flow in a cavity. Internat. *J. Numer. Methods Fluids* **5**, 561–575.

Furukawa, Y. (1982). Structures and formation mechanisms of snow polycrystals. *J. Meteorol. Soc. Jpn.* **60**, 535–547.

Goldstein, S. (1929). Forces on a solid body moving through viscous fluid. *Proc. Roy. Soc. (London)* **A123**, 216–235.

Grove, A. S., Shair, F. H., Petersen, E. E., and Acrivos, A. (1964). Experimental investigation of steady separated flow past a circular cylinder. *J. Fluid Mech.* **19**, 60.

Hadamard, J. (1911). Mouvment permanent lent d'une sphere liquid et visqueuse dans un liquid visqueux. *Compt. Rend.* **152**, 1735–1738.

Hall, W. D., and Pruppacher, H. R. (1976). The survival of ice crystals falling from cirrus clouds in subsaturated air. *J. Atmos. Sci.* **33**, 1995–2006.

Hallett, J. (1964). Experimental studies on the crystallization of supercooled water. *J. Atmos. Sci.* **21**, 671–682.

Hamielec, A. C., and Raal, J. D. (1969). Numerical studies of viscous flow around a circular cylinder. *Phys. Fluids* **12**, 11–22.

Hamielec, A. E., Hoffman, J. W., and Ross, L. (1967). *A. I. Chem. Eng. J.* **13**, 212.

Happel, J., and Brenner, H. (1965). "Low Reynolds Number Hydrodynamics," Prentice-Hall, Englewood Cliffs, NJ.

Harimaya, T. (1975). The riming property of snow crystals. *J. Meteorol. Soc. Jpn.* **53**, 384–392.

Heymsfield, A. J. (1972). Ice crystal terminal velocities. *J. Atmos. Sci.* **29**, 1348–1357.

Heymsfield, A. J. (1975). Cirrus uncinus generating cells and the evolution of cirroform clouds. Part I: Aircraft observations of the growth of ice phase. *J. Atmos. Sci.* **32**, 799–808.

Heymsfield, A. J. (1978). The characteristics of graupel particles in northeastern Colorado cumulus congestus clouds. *J. Atmos. Sci.* **35**, 284–295.

Hidy, G. M., and Brock, J. D. (1971). The dynamics of aerocolloidal systems. In "International Reviews in Aerosol Physics and Chemistry," Vol. 1. Pergamon Press, Oxford.

Hobbs, P. V. (1976). "Ice Physics." Oxford University Press.

Irbarne, J. V., and Cho, H. R. (1980). "Atmospheric Physics." D. Reidel, 212pp.

Jackson, J. D. (1974). "Classical Electrodynamics." Wiley, New York.

Jayaweera, K. O. L. F. (1971). Calculation of ice crystal growth. *J. Atmos. Sci.* **28**, 728–736.

Jayaweera, K. O. L. F., and Mason, B. J. (1965). The behavior of freely falling cylinders and cones in a viscous fluid. *J. Fluid Mech.* **22**, 709–720.

Jenson, V. G. (1959). Viscous flow round a sphere at low Reynolds numbers. *Proc. Roy. Soc. (London)* **A249**, 346–366.

Ji, W., and Wang, P. K. (1989). Numerical simulation of three dimensional unsteady viscous flow past hexagonal ice crystals in the air—Preliminary results. *Atmos. Res.* **25**, 539–557.

Ji, W., and Wang, P. K. (1991). Numerical simulation of three-dimensional unsteady viscous flow past finite cylinders in an unbounded fluid at low intermediate Reynolds numbers. *Theor. Compu. Fluid Dynam.* **3**, 43–59.

Ji, W., and Wang, P. K. (1998). On the ventilation coefficients of falling ice crystals at low-intermediate Reynolds numbers. *J. Atmos. Sci.* **56**, 829–836.

Johnson, D. E., Wang, P. K., and Straka, J. M. (1993). Numerical simulation of the 2 August 1981 CCOPE supercell storm with and without ice microphysics. *J. Appl. Meteorol.* **32**, 745–759.

Johnson, D. E., Wang, P. K., and Straka, J. M. (1994). A study of microphysical processes in the 2 August 1981 CCOPE supercell storm. *Atoms. Res.* **33**, 93–123.

Kajikawa, M. (1974). On the collection efficiency of snow crystals for cloud droplets. *J. Meteorol. Soc. Jpn.* **52**, 328–336.

Kajikawa, M., and Heymsfield, A. J. (1989). Aggregation of ice crystals in cirrus. *J. Atmos. Sci.* **46**, 3108–3121.

Kikuchi, K., and Uyeda, H. (1979). On snow crystals of spatial dendritic type. *J. Meteorol. Soc. Jpn.* **57**, 282–287.

Knight, C. A., and Knight, N. C. (1970). The falling behavior of hailstones. *J. Atmos. Sci.* **27**, 672–681.

Kobayashi, T., Furukawa, Y., Kikuchi, K., and Uyeda, H. (1976a). On twinned structures in snow crystals. *J. Crystal Growth* **32**, 233–249.

Kovasznay, L. S. G. (1949). Hot wire investigation of the wake behind cylinders at low Reynolds numbers. *Proc. Roy. Soc.* **A198**, 174–190.

Lai, K. Y., Dayan, N., and Kerker, M. (1978). Scavenging of aerosol particles by falling water drops. *J. Atmos. Sci.* **35**, 674–682.

Le Clair, B. P., Hamielec, A. E., and Pruppacher, H. R. (1970). A numerical study of the drag on a sphere at low and intermediate Reynolds numbers. *J. Atmos. Sci.* **27**, 308–315.

Lee, C. W. (1972). On the crystallographic orientation of spatial branches in natural polycrystalline snow crystals. *J. Meteorol. Soc. Jpn.* **50**, 171–180.

Leonard, B. P. (1979). Adjusted quadratic upstream algorithms for transient incompressible convection. *In* "A Collection of Technical Papers: AIAA Computational Fluid Dynamics Conference." New York. AIAA Paper 79-1649.

Leong, K. H., Beard, K. V., and Ochs, H. (1982). Laboratory measurements of particle capture by evaporating cloud drops. *J. Atmos. Sci.* **39**, 1130–1140.

Lin, Hsin-Mu, and Wang, P. K. (1997). A numerical study of microphysical processes in the 21 June 1991 Northern Taiwan mesoscale precipitation system. *Terres. Atmos. Oceanic Sci.* **8**, 385–404.

Lin, R. F. (1997). A numerical study of the evolution of nocturnal cirrus by a two-dimensional model with explicit microphysics. Ph.D. Dissertation, Dept. of Meteorology, The Pennsylvania State University.

List, R., and Schememauer, R. S. (1971). Free-fall behavior of planar snow crystals, conical graupel, and small hail. *J. Atms. Sci.* **28**, 110–115.

Liu, H. C. (1999). "A numerical study on the development of cirrus clouds." Ph.D. Thesis. Department of Atmospheric and Oceanic Sciences, University of Wisconsin-Madison.

Liu, H. C., Wang, P. K., and Schlesinger, R. E. (2001a). A Numerical Study of Cirrus Clouds. Part I: Model Description. (submitted to *J. Atmos. Sci.*)

Liu, H. C., Wang, P. K., and Schlesinger, R. E. (2001b). A Numerical Study of Cirrus Clouds. Part II: Effects of Ambient Temperature and Stability on Cirrus Evolution. (submitted to *J. Atmos. Sci.*)

Liu, H. C., Wang, P. K., and Schlesinger, R. E. (2001c). A Numerical Study of Cirrus Clouds. Part III: Effect of Radiation, Ice Microphysics and Microdynamics on Cirrus Evolution. (submitted to *J. Atmos. Sci.*)

Locatelli, J. D., and Hobbs, P. V. (1974). Fall speeds and masses of solid precipitation particles. *J. Geophys. Res.* **79**, 2185–2197.

Lorrain, P., and Carson, D. (1967). "Electromagnetic Fields and Waves." Freeman.

Magono, C., and Lee, C. W. (1966). Meteorological classification of natural snow crystals. *J. Fac. Sci. Hokkaido University, Ser. VII,* **2,** 321–335.

Magono, C., Endoh, T., Harimaya, T., and Kubota, S. (1974). A measurement of scavenging effect of falling snow crystals on the aerosol concentration. *J. Meteorol. Soc. Jpn.* **52,** 407–416.

Magradze, G. J., and Wang, P. K. (1995). A note on the closed-form mathematical description of the volume of conical hydrometeors. *Atmos. Res.* **39,** 275–278.

Manabe, S., and Strickler, R. F. (1964). Thermal equilibrium of the atmosphere with a convective adjustment. *J. Atmos. Sci.* **21,** 361–385.

Marshall, J. S., and Palmer, W. M. (1948). The distribution of raindrops with size. *J. Meteorol.* **5,** 165–166.

Martin, J. J., Wang, P. K., and Pruppacher, H. R. (1980a). On the efficiency with which aerosol particles of radius larger than 0.1 micron are collected by simple ice plates. *Pure Appl. Geophys.* **118,** 1109–1129

Martin, J. J., Wang, P. K., and Pruppacher, H. R. (1980b). A theoretical determination of the efficiency with which aerosol particles are collected by simple ice plates. *J. Atmos. Sci.* **37,** 1628–1638.

Martin, J. J., Wang, P. K., Pruppacher, H. R., and Pitter, R. L. (1981). A numerical study of the effect of electric charges on the efficiency with which planar ice crystals collect supercooled water drops. *J. Atmos. Sci.* **38,** 2462–2469.

Masliyah, J. H., and Epstein, N. (1971). Numerical study of steady flow past spheroids. *J. Fluid Mech.* **44,** 493–512.

Mason, B. J. (1971). "The Physics of Clouds," 2nd Ed., Oxford University Press.

McDonald, J. E. (1963). Use of electrostatic analogy in studies of ice crystal growth. *Z. Angew. Math. Phys.* **14,** 610.

Meyers, M. P., DeMott, P. J., and Cotton, W. R. (1992). New primary ice nucleation parameterizations in an explicit cloud model. *J. Appl. Meteorol.* **31,** 708–721.

Miller, N. L. (1988). A Theoretical Investigation of the Collection of Aerosol Particles by Falling Ice Crystals. PhD Thesis, Department of Meteorology, University of Wisconsin-Madison, 205 pp.

Miller, N. L., and Wang, P. K. (1989). A theoretical determination of the efficiency with which aerosol particles are collected by falling columnar ice crystals. *J. Atmos. Sci.* **46,** 1656–1663.

Mitchell, D. L. (1996). Use of mass- and area-dimensional power laws for determining precipitation particle terminal velocities. *J. Atmos. Sci.* **53,** 1710–1723.

Mitchell, D. L. (1998). Parameterizing the extinction and absorption in ice clouds: A process oriented approach. Conf. on Light Scattering by Non-spherical Particles: Theory, Measurements and Applications. 29 September–1 October, 1998, New York.

Mitchell, D. L., Zhang, R., and Pitter, R. L. (1990). Mass-dimensional relationships for ice particle size spectra and radiative properties of cirrus clouds. Part II: Dependence of absorption and extinction on ice crystal morphology. *J. Atmos. Sci.* **81,** 817–832.

Mitra, S. K., Barth, S., and Pruppacher, H. R. (1990). A laboratory study of the scavenging of SO2 by snow crystals. *Atmos. Environ.* **24,** 2307–2312.

Morse, P. M., and Feshbach, H. (1953). "Methods of Theoretical Physics," Vol. 2. McGraw-Hill, New York.

Murakami, M., Kikuchi, K., and Magono, C. (1985a). Experiments on aerosol scavenging by snow crystals. Part I: Collection efficiencies of uncharged snow crystals for micron and submicron particles. *J. Meteorol. Soc. Jpn. Ser. II,* **63,** 119–129.

Murakami, M., Kikuchi, K., and Magono, C. (1985a). Experiments on aerosol scavenging by snow crystals. Part II: Attachment rate of 0.1 micron diameter particles to stationary snow crystals. *J. Meteorol. Soc. Jpn. Ser. II,* **63,** 130–135.

Needham, J., and Lu, G. D. (1961). The earliest snow crystal observations. *Weather* **XVI,** 319–327.

Ono, A. (1969). The shape and riming properties of ice crystals in natural clouds. *J. Atmos. Sci.* **26**, 138–147.

Oseen, C. W. (1910). Uber die Stokessche Formel und uber die verwandte Aufgabe in der Hydrodynamik. *Arkiv Mat., Astron., Fysik* **6**(29).

Parungo, F. (1995). Ice crystals in high clouds and contrails. *Atmos. Res.* **38**, 249–262.

Pasternak, I. S., and Gauvin, W. H. (1960). Turbulent heat and mass transfer from stationary particles. *Can. J. Chem. Eng.* **38**, 35–42.

Peyret, R., and Taylor, T. D. (1983). Computational Methods for Fluid Flow, Springer-Verlag, Berlin.

Pitter, R. L. (1977). A re-examination of the riming on thin ice plates. *J. Atmos. Sci.* **34**, 684–685.

Pitter, R. L., and Pruppacher, H. R. (1974). A numerical investigation of collision efficiencies of simple ice plates colliding with supercooled water drops. *J. Atmos. Sci.* **31**, 551–559.

Pitter, R. L., Pruppacher, H. R., and Hamielec, A. E. (1973). A numerical study of viscous flow past thin oblate spheroid at low and intermediate Reynolds numbers. *J. Atmos. Sci.* **30**, 125–134.

Pitter, R. L., Pruppacher, H. R., and Hamielec, A. E. (1974). A numerical study of the effect of forced convection on mass transfer from a thin oblate spheroid in air. *J. Atmos. Sci.* **31**, 1058–1066.

Podzimek, J. (1987). Direct scavenging and induced transport of atmospheric aerosol by falling snow and ice crystals. *In* "Seasonal Snow Cover: Physics, Chemistry, and Hydrology" (H. G. Jones and W. J. Orville-Thomas, eds.). Reidel, New York, pp. 209–244.

Prodi, F. (1976). Scavenging of aerosol particles by growing ice crystals. Preprints. Int. Conf. Cloud Physics. 30 July 1976, Boulder, CO, pp. 70–75.

Pruppacher, H. R., and Klett, J. D. (1978). "Microphysics of Clouds and Precipitation." D. Reidel, New York.

Pruppacher, H. R., and Klett, J. D. (1997). "Microphysics of Clouds and Precipitation." Kluwer Academic Publishers, Amsterdam.

Pruppacher, H. R., and Pitter, R. L. (1971). A semi-empirical determination of the shape of cloud and rain drops. *J. Atmos. Sci.* **28**, 86–94.

Pruppacher, H. R., Le Clair, B. P., and Hamielec, A. E. (1970). Some relations between the drag and flow pattern of viscous flow past a sphere and a cylinder in low and intermediate Reynolds numbers. *J. Fluid Mech.* **44**, 781–790.

Pruppacher, H. R., Rasmussen, R., Walcek, C., and Wang, P. K. (1982). A wind tunnel investigation of the shape of uncharged raindrops in the presence of an external vertical electric field. Proc. 2nd Int. Colloq. on Drops and Bubbles, Monterey, CA, pp. 239–243.

Ramanathan, V., Pitcher, E. J., Malone, R. C., and Blackmon, M. L. (1983). The response of a spectral general circulation model to refinements in radiative processes. *J. Atmos. Sci.* **40**, 605.

Ramaswamy, V., and Ramanathan, V. (1989). Solar absorption by cirrus clouds and the maintenance of the tropical upper troposphere thermal structure. *J. Atmos. Sci.* **46**, 2293–2310.

Randall, D. A., Harshvardhan, and Dazlich, D. A. (1989). Interactions among radiation, convection, and large scale dynamics in a general circulation model. *J. Atmos. Sci.* **46**, 1943–1970.

Reinking, R. F. (1979). The onset and early growth of snow crystals by accretion of droplets. *J. Atmos. Sci.* **36**, 870–881.

Richards, C. N., and Dawson, G. A. (1971). The hydrodynamic instability of water drops falling at terminal velocity in vertical electric fields. *J. Geophys. Res.* **76**, 3445–3455.

Rimon, Y., and Lugt, H. J. (1969). Laminar flows past oblate spheroids of various thicknesses. *Phys. Fluids* **12**, 2465–2472.

Rybczinski, W. (1911). *Bull. Acad. Caracovie, Ser. A.,* p. 40 [referred to in Lamb, H. (1932). "Hydrodynamics," Cambridge University Press.

Sasyo, Y. (1971). Study of the formation of precipitation by the aggregation of snow particles and the accretion of cloud droplets on snowflakes. *Pap. Meteorol. Geophys.* **22**, 69–142.

Sasyo, Y., and Tokuue, H. (1973). The collection efficiency of simulated snow particles for water droplets (preliminary report). *Pap. Meteorol. Geophys.* **24**, 1–12.

Sauter, D. P., and Wang, P. K. (1989). An experimental study of the scavenging of aerosol particles by natural snow crystals. *J. Atmos. Sci.* **46**, 1650–1655.

Schlamp, R. J., Pruppacher, H. R., and Hamielec, A. E. (1975). A numerical investigation of the efficiency with which simple columnar ice crystals collide with supercooled water drops. *J. Atmos. Sci.* **32**, 2330–2337.

Schlamp, R. J., Grover, S. N., Pruppacher, H. R., and Hamielec, A. E. (1976). A numerical investigation of the electric charges and vertical external electric fields on the collision efficiency. *J. Atmos. Sci.* **33**, 1747–1755.

Seinfeld, J. H., and Pandis, S. N. (1997). "Atmospheric Chemistry and Physics," John Wiley & Sons, New York.

Slinn, W. G. N., and Hale, J. M. (1971). A re-evaluation of the role of thermophoresis as a mechanism of in- and below-cloud scavenging. *J. Atmos. Sci.* **28**, 1465–1471.

Smythe, W. R. (1956). Charged right circular cylinders. *J. Appl. Phys.* **27**, 917–920.

Smythe, W. R. (1962). Charged right circular cylinders. *J. Appl. Phys.* **33**, 2966–2967.

Song, N., and Lamb, D. (1992). Aerosol scavenging by ice in supercooled clouds. *In* "Precipitation Scavenging and Atmosphere–Surface Exchange," Vol. I. pp. 63–74.

Srivastava, R. C., and Coen, J. L. (1992). New explicit equations for the accurate calculations of the growth and evaporation of hydrometeors by the diffusion of water vapor. *J. Atmos. Sci.* **49**, 1643–1651.

Starr, D. O'C., and Cox, S. K. (1985a). Cirrus clouds, I: A cirrus cloud model. *J. Atmos. Sci.* **42**, 2663.

Starr, D. O'C., and Cox, S. K. (1985b). Cirrus clouds, II: Numerical experiments on the formation and maintenance of cirrus. *J. Atmos. Sci.* **42**, 2682.

Straka, J. M. (1989). Hail growth in a highly glaciated Central High Plains multi-cellular hailstorm. Ph.D. Thesis, Department of Meteorology, University of Wisconsin-Madison.

Sturnilo, O., Mugnai, A., and Prodi, F. (1995). A numerical sensitivity study on the backscattering at 35.8 GHz from precipitation-sized hydrometeors. *Radio Sci.* **30**, 903–919.

Takahashi, T. (1973). Measurement of electric charge on cloud drops, drizzle drops and rain drops. *Rev. Geophys. Space Phys.* **11**, 903–924.

Takahashi, T., Endoh, T., Wakahama, G., and Fukuta, N. (1991). Vapor diffusional growth of free-falling snow crystals beween −3 and −23C. *J. Meteorol. Soc. Jpn.* **69**, 15–30.

Takami, H., and Keller, H. B. (1969). Steady two-dimensional viscous flow of an incompressible fluid past a circular cylinder. *Phys. Fluid Suppl.* II, pp. 51–56.

Thom, A. (1933). The flow past circular cylinders at low speeds. *Proc. Roy. Soc.* **A141**, 651.

Thorpe, A. D., and Mason, B. J. (1966). The evaporation of ice spheres and ice crystals. *Brit. J. Appl. Phys.* **17**, 541–548.

Tremback, C. J. Powell, Cotton, W. R., and Pielke, R. A. (1987). The forward-in-time upstream advection scheme: Extension to higher order. *Mon. Wea. Rev.* **115**, 540–555.

Verlinde, J. P., Flatau, J., and Cotton, W. R. (1990). Analytical solutions to the collection growth equation: Comparison with approximate methods and application to cloud microphysics parameterization schemes. *J. Atmos. Sci.* **47**, 2871–2880.

Wang, P. K. (1982). Mathematical description of the shape of conical hydrometeors. *J. Atmos. Sci.* **39**, 2615–2622.

Wang, P. K. (1983a). On the definition of collision efficiency of atmospheric particles. *J. Atmos. Sci.* **40**, 1051–1052.

Wang, P. K. (1983b). Collection of aerosol particles by conducting spheres in an external electric field-continuum regime approximation. *J. Coll. Interf. Sci.* **94**, 301–318.

Wang, P. K. (1985). A convective diffusion model for the scavenging of submicron particles by snow crystals of arbitrary shapes. *J. de Rech. Atmos.* **19**, 185–191.

Wang, P. K. (1987). Two dimensional characterization of polygonally symmetric particles. *J. Colloid Interf. Sci.* **117**, 271–281.

Wang, P. K. (1997). Characterization of ice particles in clouds by simple mathematical expressions based on successive modification of simple shapes. *J. Atmos. Sci.* **54**, 2035–2041.

Wang, P. K. (1999). Three-dimensional representations of hexagonal ice crystals and hail particles of elliptical cross-sections. *J. Atmos. Sci.* **56**, 1089–1093.

Wang, P. K., and Denzer, S. M. (1983). Mathematical description of the shape of plane hexagonal snow crystals. *J. Atmos. Sci.* **40**, 1024–1028.

Wang, P. K., and Jaroszczyk, T. (1991). The grazing collision angle of aerosol particles colliding with infinitely long circular cylinders. *Aerosol Sci. Tech.* **15**, 149–155.

Wang, P. K., and Ji, W. (1992). A numerical study of the diffusional growth and riming rates of ice crystals in clouds. Preprint Vol. 11th Int. Cloud Physics Conf. Aug. 11–17, 1992, Montreal, Canada.

Wang, P. K., and Ji, W. (1997). Simulation of three-dimensional unsteady flow past ice crystals. *J. Atmos. Sci.* **54**, 2261–2274.

Wang, P. K., and Ji, W. (2000). Collision Efficiencies of Ice Crystals at Low-Intermediate Reynolds Numbers Colliding with Supercooled Cloud Droplets: A Numerical Study. *J. Atmos. Sci.* **57**, 1001–1009.

Wang, P. K., and Lin, H. (1995). Comparison between the collection efficiency of aerosol particles by individual water droplets and ice crystals in a subsaturated atmosphere. *Atmos. Res.* **38**, 381–390.

Wang, P. K., and Pruppacher, H. R. (1977). An experimental determination of the efficiency with which aerosol particles are collected by water drops in subsaturated air. *J. Atmos. Sci.* **34**, 1664–1669.

Wang, P. K., and Pruppacher, H. R. (1980a). The effect of an external electric field on the scavenging of aerosol particles by clouds and small rain drops. *J. Colloid Interf. Sci.* **75**, 286–297.

Wang, P. K., and Pruppacher, H. R. (1980b). On the efficiency with which aerosol particles of radius less than one micron are collected by columnar ice crystals. *Pure Appl. Geophys.* **118**, 1090–1108.

Wang, P. K., Chuang, C. H., and Miller, N. L. (1985). Electrostatic, temperature, and vapor density fields surrounding stationary columnar ice crystals. *J. Atmos. Sci.* **42**, 2371–2379.

Wang, P. K., Greenwald, T. J., and Wang, J. (1987). A three parameter representation of the shape and size distributions of hailstones—A case study. *J. Atmos. Sci.* **44**, 1062–1070.

Wang, P. K., Grover, S. N., and Pruppacher, H. R. (1978). On the effect of electric charges on the scavenging of aerosol particles by cloud and small rain drops. *J. Atmos. Sci.* **35**, 1735–1743.

Warren, S. G. (1984). Optical constants of ice from ultraviolet to microwave. *Appl. Optics.* **23**, 1206–1225.

Warren, S. G., Hahn, C. J., London, J., Chervin, R. M., and Jenne, R. L. (1986). Global distribution of tital cloud cover and cloud type amounts over land. NCAR Tech. Note NCAR/TN-273+STR, National Center for Atmospheric Research, Boulder, CO.

Warren, S. G., Hahn, C. J., London, J., Chervin, R. M., and Jenne, R. L. (1988). Global distribution of tital cloud cover and cloud type amounts over land. NCAR Tech. Note NCAR/TN-317+STR, National Center for Atmospheric Research, Boulder, CO.

Wilkins, R. I., and Auer, A. H. (1970). Riming properties of hexagonal ice crystals. *In* Preprints, Conf. Cloud Physics., Fort Collins, Aug. 24–27, 1970, pp. 81–82.

Willmarth, W. W., Hawk, N. E., and Harvey, R. L. (1964). Steady and unsteady motions and wakes of freely falling disks. *Phys. Fluids* **7**, 197–208.

Wylie, D. P., and Menzel, W. P. (1989). Two years of cloud cover statistics using VAS. *J. Clim. Appl. Meteorol.* **2**, 380–392.

Yee, H. C. (1987). Upwind and Symmetric Shock-Capturing Schemes. NASA Technical Memorandum 89464.

Yih, Chia-Shun (1969). "Fluid Mechanics." McGraw-Hill, New York.

INDEX